TROUBLED WATERS

Confronting the Water Crisis
in Australia's Cities

TROUBLED WATERS

Confronting the Water Crisis in Australia's Cities

Edited by Patrick Troy

ANU

THE AUSTRALIAN NATIONAL UNIVERSITY

E PRESS

ANU E PRESS

Published by ANU E Press
The Australian National University
Canberra ACT 0200, Australia
Email: anuepress@anu.edu.au
This title is also available online at: http://epress.anu.edu.au/troubled_waters_citation.html

National Library of Australia
Cataloguing-in-Publication entry

National Library of Australia Cataloguing-in-Publication entry

Title: Troubled waters : confronting the water crisis in
                  Australia's cities / editor, Patrick Troy.
ISBN: 9781921313837 (pbk.)
                  9781921313844 (pdf.)
Notes: Includes index.
                  Bibliography.
Subjects: Water consumption--Australia
                  Water conservation--Australia
                  Water resources development--Australia
                  Dwellings--Australia--Energy consumption.
                  Climatic changes--Environmental aspects--Australia.
Other Authors/Contributors:
                  Troy, Patrick N. (Patrick Nicol), 1936-
Dewey Number: 363.610994

Cover design by ANU E Press
Cover images by Silvia, sourced from:
http://www.stockvault.net/Sea_water_g13-Drop_of_water_p9502.html
http://www.stockvault.net/Food_Drink_g16-Water_bubbles_p9504.html

Contents

Contributors

Patrick Troy AO is Emeritus Professor and Visiting Fellow in the Fenner School of Environment and Society, Australian National University <patrick.troy@anu.edu.au>

Tony Dingle is Professor of Economics, Monash University. <Tony.Dingle@BusEco.monash.edu.au>

Peter Spearritt is Professor of History, Centre for Applied History and Heritage Studies, University of Queensland. <p.spearritt@uq.edu.au>

Graeme Davison is Sir John Monash Distinguished Professor, School of Historical Studies, Monash University. <Graeme.Davison@arts.monash.edu.au>

Lesley Head is Professor, School of Earth & Environmental Sciences and GeoQuEST Research Centre University of Wollongong. <lhead@uow.edu.au>

Stephen Dovers is Professor, Fenner School of Environment and Society. Australian National University. <stephen.dovers@anu.edu.au>

Geoffrey J. Syme is Research Program Leader of Society, Economy and Policy, CSIRO Land and Water. <Geoff.Syme@csiro.au>

Janice Gray is Senior Lecturer, School of Law, University of New South Wales. <j.gray@unsw.edu.au>

Alex Gardener is Associate Professor, School of Law, Law University of Western Australia. <agardner@law.uwa.edu.au>

Lee Godden is Associate Professor, Faculty of Law, and Director, Office for Environment Programs, University of Melbourne. <l.godden@unimelb.edu.au>

Acknowledgments

The conversations, Symposium, National Workshop and explorations on which this book is based were sponsored by the Academy of Social Sciences in Australia and the Australian National University's Water Initiative.

Individual authors also acknowledge assistance as follows:

Patrick Troy wishes to acknowledge the contributions of the five anonymous referees; Jim Walter for his guidance; and Phil Greaves for his patience and support.

Graeme Davison acknowledges the assistance of Frank Bongiorno, Miles Lewis, Richard Overell and Kimberley Webber in obtaining source material for this essay. He also thanks Katherine Rogerson of the ACP Library for permission to reproduce the cover of *The Australian Women's Weekly*, 14th April 1951, and illustrations; Sandra Camden-Bermingham of Electrolux Home Products for permission to reproduce the Hoover advertisment; *Australia Home Beautiful* for illustrations and Victa for permission to reproduce the illustration.

Lesley Head wishes to acknowledge that The Backyard Project was funded by the Australian Research Council (DP 0211327).

Janice Gray and Alex Gardener would particularly like to thank Dr Michael Keating, Chairperson, IPART; Pamela Soon, Solicitor, IPART; Luke Woodward, Partner, Gilbert + Tobin Lawyers; Sally Walkom, Senior Policy Advisor, Sydney Water; and Ian Waters, Corporate Secretary and Solicitor, Sydney Water Corporation, whose assistance was invaluable.

All errors, omissions and misinterpretations remain those of the authors.

Introduction: The water services problem

Patrick Troy

Despite the urgency with which Australian cities now face the problem of inadequate water supplies, and despite the impact of recent patterns of changes in climate on those supplies, the roots of the water problem are deeply historical and can only be addressed by accounting for intersecting technological, cultural, economic and political factors. Together these factors have entrenched a path dependency in the way water services are supplied and attitudes towards them that must be thoroughly questioned if the current crisis is to be understood and addressed. This book summarises these intersections as a preliminary to the consideration of alternative methods of ensuring a sustainable and appropriate mix of the supply of water services for Australian cities in the future.

In Chapter 1, Tony Dingle points out that from the standpoint of today, urban water supply and sewerage systems in Australia appear to have changed little in their technological essentials for extended periods of time. A water-supply authority harvests rainwater or stream flow, stores it, treats it (if required) so that it is potable, and then reticulates it into homes and businesses. The liquid and solid wastes generated there are piped out again, treated to some degree, before being emptied into the ocean.

This has usually been explained as an example of path dependence. The technology is expensive and disruptive to install. Once in place it limits future development paths so that it is less costly to expand an existing system as the number of consumers and per-capita requirements grow, rather than replace it with something different. Consumers also are not required to change their usage habits. This generates a set of obstacles both in supply and usage to the adoption of new technologies that differ significantly from what is currently in place.

There have, however, been major transitions in the past where cities have replaced existing methods of supply with new and different technologies. The most obvious of these have been the shifts from water delivery by water cart or on-site rainfall collection to reticulation; and from nightsoil collection to piped, water-borne collection by means of flush toilets. These transitions were made by most cities in the developing world between the mid nineteenth and the mid twentieth century. Typically, they were climactic events provoked by widespread concerns, usually about threats to public health. They usually involved an assessment of the alternative technologies available at the time.

In examining these transitions, their timing and the factors which provoked them, Dingle raises questions about what might constitute necessary conditions for radical change in today's networked cities. In the light of previous transitions, are they now approaching the critical conditions in relation to water supply and sewerage provision that could push them onto different trajectories in relation to present and future sources of water as well as patterns of water usage?

In Chapter 2, Peter Spearritt provides illustrations of the way the rigidities in approaches to the provision of water services and short-term political considerations have influenced the investment in water-services infrastructure in Southeast Queensland (SEQ). He discusses how the Queensland Government reacted to the recent drought and argues that the explanation of the crisis in SEQ lies in an analysis of infrastructure quick-fixes popular with engineers, a remarkable lack of accountability in the water bureaucracies and untold arrogance in the electricity authorities. Spearritt claims that successive bureaucrats and Ministers ignored the warning signs, sounded as early as 1997 by experts in the Department for Natural Resources and Mines. The Government embarked instead on a series of 'supply side' projects although the creation of the Queensland Water Commission in 2006 was followed by the introduction of regulations which demanded severe reductions in domestic water consumption.

In Chapter 3, Graeme Davison explores the social history of our changing water-consumption habits and behaviour. He argues that the increase in consumption was brought about by the twin concerns of health and morality, reminding us of the strong link between Protestant morality and modern habits of cleanliness. Davison explores further the connection between the views expressed by the reformer Chadwick and the development of the sewerage systems in Australian cities as outlined by Dingle in Chapter 1.

He discusses the importance of fashion and technological developments in bathing and personal cleanliness and in washing, as well as in the factors behind the great increase in water consumption outside the house, in the garden and recreation facilities such as swimming pools.

In Chapter 4, Lesley Head argues that although the availability or otherwise of water has always underpinned the human settlement of Australia, the fundamental nature of our relationship with water cycles in and out of public consciousness. Recent drought has returned the issue to the top of public agendas. In debates over dam expansion, recycling sewage, desalination, watering lawns and washing cars, a key motif that swirls around is that Australians need a 'culture change' in relation to water. It is said that we need to change attitudes of profligacy, developed in the well-watered ancestral lands of northwest Europe, and attune both attitude and practice to the realities of living on the driest inhabited continent on Earth.

But what would constitute such a culture change, and how would we recognise it? She takes an ethnographic approach to the question of urban water use through the lens of the backyard garden, drawing on interview material from a broader study to examine the ways in which people think about and use water. In arguing that there is a significant cultural shift occurring, she does not discuss the actual levels of water consumption but offers a complementary perspective that seeks to understand everyday practices and habits, and the processes that reinforce or change them.

Lesley's argument contrasts with and extends other studies that have emphasised the perceived separation between the modern home and the networks of production that sustain it. She argues that it is in the relationship between house and garden that people see, understand and participate in the network of water storage and distribution. Their active engagement with these processes enhances their capacity to manage and reduce consumption.

In Chapter 5, Stephen Dovers seeks to connect discussion of human behaviours around water not to taps, toilets and timing showers, or dams and desal plants, as much discussion (very usefully) does, but to the policy processes and instruments, institutional and governance systems, and household realities that shape human and organisational behaviours toward water in a modern society and economy. The focus is on urban water, but the discussion necessarily travels to rural water and issues such as energy that cannot easily be separated from water. The paper is a series of linked discussions on issues that surround more singular policy debates around water, hinged on the proposition that water policy is better constructed as being about far more than just water, and where the prospects for behavioural and institutional change become both more complicated and realistic.

In Chapter 6, Geoff Symes argues that although there is now a substantial literature defining and encouraging sustainable urban water management the responses of urban water utilities to changed demand and supply options have tended to focus on technological solutions, new sources and well-worn approaches to demand management. While there is increasing interest in water-sensitive urban design, whole-of-lifecycle economic consideration and the incorporation of externalities into pricing and cost-benefit analyses, there are significant areas of sustainability that have received scant attention. These neglected areas tend to relate to the difficulty in creating, as opposed to promoting, the concept of sustainability that includes social and cultural assessment, integrated response and key institutional issues in achieving adaptive learning.

Recently, however, there have been attitude changes towards supporting stronger approaches to sustainable water-resources management. There is also a greater appreciation that holistic approaches will have to be taken as the issues associated with metropolitan growth and climate change have become more evident.

Many in the community realise that there are important value judgments that will have to be addressed to establish whether the status quo should be maintained. Alternatively, if strategic social and cultural goals are to be achieved the community are willing to engage in examining what novel institutional structures should be seriously considered to attain them.

Research has shown that these drivers include judgments on issues of fairness in allocation, acceptable risk and uncertainty, trust in both government and its agencies, and perceived wellbeing from alternative levels of service. Emotion is also a significant driver of community decision-making. These judgments are underpinned by perceptions of professional roles and knowledge and how they are incorporated in public discussion.

Geoff discusses these issues and the prospects of, and justification for, change in decision-making in urban water management in terms of examples of community water culture in relation to alternative delivery systems and inter-regional transfer of water resources.

In Chapter 7, Janice Gray and Alex Gardner pose the question of how to provide new third-party (and usually private-sector) access to old public-sector infrastructure in order to make better use of these valuable water resources. They also explore the associated issues of how to maintain health standards and societal protections in the face of private-sector involvement in the supply of fundamental life services.

Providing for private-sector access to wastewater infrastructure and facilitating private-sector wastewater services requires sophisticated levels of science and technology. It also requires the development of an appropriate legislative framework to regulate the private-sector access and services. Regulatory wastewater regimes need to operate in tandem with the broader legal framework for water services provision, including the economic regulatory bodies, at both the State and Commonwealth levels.

They explore the opportunities that may exist for third-party access to public infrastructure and, accordingly, sketch the present institutional frameworks for water and wastewater management throughout Australia. They then review the *Water Industry Competition Act* 2006 (NSW) as an example of state-based industry-specific legislation incorporating a third-party access regime. They tease out some points of interest and potential concerns associated with third-party access regimes.

In Chapter 8, Lee Godden argues that property, far from being the settled and determinate concept that people ascribe to the word, remains a contested site for defining 'rights' — and for articulating obligations. This chapter traces the changing conceptions of property in Australian law by reference to whether property is an appropriate descriptor for defining and managing 'entitlements' to water in an urban context. Debates about the utility of property concepts and market mechanisms in achieving the goals of sustainable water use and economic efficiency have, to date, largely focused upon the rural sector. The social-justice dimensions of these approaches, initiated under the Council of Australian Government reforms to water laws and National Competition Policy reforms, is now an emerging issue. Many of the questions about the balance between private rights and the public interest that arise in this context have similar resonances in an urban setting.

In urban areas, though, the issues of defining the public interest and accountability, vis-à-vis the 'rights' of water consumers, take on particular dimensions in the light of the regulatory changes that have occurred in the water-supply and 'retail' sectors. The moves to deregulate urban water authorities have created semi-corporatised models of governance across these sectors. Lee examines the changes in urban water regulation, assessing whether various models embraced under public–private partnerships and de-regulated structures can promote sustainable use of urban water and provide an effective means of 'balancing' the various interests. She explores new forms of property-based measures, such as 'offsets' trading or water 'credits', and their potential to deliver sustainable, long-term urban water use.

Arguably, there is a need to reiterate and redefine the long-term responsibilities of governments, individual water users and the community as a whole in regard to urban water. Such a long-term view needs to be accompanied by an articulation of a holistic 'responsibilities' spectrum rather than creating an artificial division into public and private spheres.

In the Conclusion, Patrick Troy argues that there is a need to find a new way of encouraging residents to take more responsibility for their water services and that central to this is a reconsideration of the Chadwickian solution of the management of human body wastes. He argues that Australian cities have reached the point where there are no further 'natural' water resources available to be exploited but that by acknowledging that residents have an inalienable right to potable water and that protection of public health remains the highest priority it is feasible to develop water services that do not increase environmental stresses.

Chapter 1

The life and times of the Chadwickian solution

Tony Dingle

This chapter provides a thumbnail sketch of the historical evolution of the water-supply and sewage-disposal systems currently in use in Australia's cities. We can begin with the inescapable physiological fact that, as humans, we need to drink a certain volume of water each day to sustain life and we produce a modest quantity of faeces and urine as wastes of our body metabolism. Were we Robinson Crusoe, or any other member of the animal kingdom, we could drink from the pure limpid stream and vacate wherever the fancy took us, confident that natural processes would recycle the nutrients. That option, however, has not been available to us for a very long time, for living in groups has required different strategies and the larger the group the more challenging the problem.

This became painfully obvious as cities began to grow with unprecedented rapidity from the beginning of the nineteenth century. As they did so, they also began killing off their residents at an alarming rate so that the death rates in these places rapidly climbed far above those of smaller towns and the countryside as a consequence of the prevalence of infectious diseases, both endemic and epidemic. The former killed more people, but the latter were more feared by city-dwellers for the suddenness of their appearance and the savagery of their impact. In the cities, death rates climbed way above birth rates, so that growth was sustained only because of the waves of migrants who continued to flood in seeking to improve their economic circumstances (Lampard 1973; Rosen 1973).

In the UK and elsewhere, physicians and others struggled to understand cause and effect. The miasma theory — that bad smells transmitted disease — proved, eventually, to be an incorrect diagnosis, but whenever it led to the cleaning up of rubbish it probably had beneficial outcomes. Under the influence of this theory, sanitary reformer Edwin Chadwick produced his masterly *Report on the Sanitary Condition of the Labouring Population of Great Britain* in 1842. He argued that it was the accumulation of filth in the cities that caused sanitary problems and his solution was the provision of a pure, piped, water supply to the towns and a water-flushed network of sewers to remove body wastes. In the calculus of his times he explained:

That the chief obstacles to the immediate removal of decomposing refuse of towns and habitations have been the expense of the hand labour and cartage requisite for the purpose.

That this expense may be reduced to one-twentieth or to one-thirtieth, or rendered inconsiderable, by the use of water and self-acting means of removal by improved and cheaper sewers and drains.

That refuse when thus held in suspension in water may be most cheaply and innoxiously conveyed to any distance out of the towns, and also in the best form for productive use, and that the loss and injury by the pollution of natural streams may be avoided.

That for all these purposes, as well as for domestic use, better supplies of water are absolutely necessary. (Chadwick 1965: 423–4)

Chadwick's solution has been almost universally adopted by cities ever since and is still being installed. Hamburg was perhaps the first large town to construct such a complete sewer system in 1843, after a big fire damaged much of the city (Derry and Williams: 426–7). The sewers were flushed weekly with river water. Paris and London, with Bazalgette's great scheme of interceptor sewers, moved in the same direction in the 1850s and 1860s. In Australia, Sydney began building sewers in the 1860s but constructed a much larger network in the 1880s. Melbourne, then a larger city, did not start building until 1892. Adelaide began in 1879, while Brisbane, which was roughly the same size as Adelaide, did not make its first connections until 1923. These differences in timing are intriguing and require investigation in a comparative framework.

There were two components to Chadwick's solution, pure piped water and water-borne sewerage collection. As more cities adopted this approach, there was an increasing focus on improving the component parts of the system so that major improvements were made to the technologies required to implement Chadwick's concept, especially during the second half of the nineteenth century. Attention was drawn to those points where improved technologies and greater knowledge would have a beneficial impact. So, for example, as larger cities sought to guarantee larger supplies of water, bigger, stronger dams were built to impound larger storages. To this time, dam building had been entirely empirical, but applied science began to make a significant contribution to the design of dams from the mid nineteenth century onwards. W. M. J. Rankine, professor of Civil Engineering at Glasgow University, worked on the properties of loose earth, thereby improving the construction of earthen dams, while the French engineer de Sazilly improved the understanding of internal stresses in masonry dams. Much larger dams of both kinds were built as the century progressed, mainly for urban water supplies (Bruce 1 1968: 556–8; Bruce 2 1968: 1368). There were some serious miscalculations — in 1864, the Dale Dyke dam near Sheffield burst, killing more than 200 people — but by the end of the nineteenth century it was common for large towns to draw their water supplies from dams a great distance away, the water travelling along masonry-lined aqueducts or, increasingly, through cast-iron pipes to its destination.

Pure water supplies could only be achieved gradually as the foundations of the new science of bacteriology were laid and understandings of disease transmission improved. London doctor John Snow made a crucial breakthrough in 1849 in understanding the most horrifying curse of British cities, cholera, when he found that it was transmitted through water supplies contaminated with the faeces of cholera victims (Wohl: 124–5). The germ theory of disease, pioneered by Louis Pasteur and others, took over from the miasma theory as micro-organisms of particular diseases were identified under the microscope. By 1880 Carl Eberth had identified the typhoid bacillus and it was realised that typhoid was transmitted by means of contact with the faeces or vomit of an infected person.

Methods of water treatment were improved significantly during the nineteenth century. Sand filters trapped suspended solids and the slow sand filter also proved to be an effective barrier to the germs of cholera and typhoid, although this was only gradually realised once advances in bacteriology had been achieved. By the end of the century it also became practicable to disinfect water by using chlorine and it began to be used on city water supplies, first at Middlekerke in Belgium in 1902; but there was significant public opposition to this practice in some cities (Bruce 2 1968: 1373–81).

With waste disposal there was a similar cluster of major technological advances which greatly improved the performance of the system, but these came somewhat later than advances in water supply, partly because they were often stimulated by the problems generated by an increased water supply and partly because they could usually only be operated on a city-wide scale and were consequently expensive and disruptive to construct and install. Sewers had existed for many centuries, both as surface gutters and as underground pipes or tunnels of some size, but their main function had been to transmit rainwater to the nearest river in order to prevent flooding of the increasingly impervious surfaces of expanding cities. These sewers, and the rivers into which they flowed, were placed under greater pressure as improved systems of water supply delivered far greater amounts for use in the home. Much of this found its way into gutters and drains once it had been used and dirtied.

The water closet became the major problem for existing methods of waste disposal. The concept of a water closet had first been developed by Sir John Harrington in 1596. In 1778, Joseph Bramah patented one with two valves that slowly came into use. A trap containing a water seal to keep smells out of the toilet was developed in 1782, and in the nineteenth century improved models were marketed by Thomas Crapper of Chelsea and many others who also added their own piecemeal improvements (Derry and Williams 1960: 425–6). In towns with improved water supplies, the wealthy increasingly installed water closets. These still emptied into cess pits or into sewers, sometimes legally and sometimes illegally. As a consequence, far more contaminated water was being trapped in the city: old sewers were not designed to flush out suspended solids and much accumulated below city streets and residences, increasing health risks through the increased likelihood of the spread of infectious diseases. The technologies of piped water and the water closet were leading, in combination, to the breakdown of existing methods of waste disposal (Tarr 1988: 162–3). They were also massively increasing river pollution.

Such outcomes hastened the construction of water-borne sewerage systems to cope with these difficulties. Once cities did this, the water closet was transformed from a major polluter to the centrepiece of the new method of waste disposal for it effortlessly flushed body wastes out beyond city boundaries as Chadwick's solution required. As a consequence, an eminent historian of sanitary reform, M. W. Flinn, has even suggested that the water closet 'may well have been in the long term the most life-saving invention of all time' (Chadwick 1965: 9).

Other advances came along with better water closets. Ovoid or egg-shaped pipes were developed to increase the scouring capabilities of sewer networks during times of low sewage flow. Flow rates and the gradients at which pipes were laid were calculated with increasing accuracy. Bazalgette's system in London was built to maintain an average speed of one-and-a-half miles per hour when the sewers were half full (Derry and Williams 1960: 427). Bricks were used to construct the larger sewers but by the end of the nineteenth century concrete was increasingly used. Large, slow-acting steam pumps were developed to pump water and sewage wherever it was not possible to use gravity. A major issue for any city was whether to build a combined or a separate system; that is, would the sewers be built large enough to cope with both sewage and stormwater, as they were in London and Paris, or would the two be channelled into separate systems? The advantage of the latter was that the sewers could be made far smaller for a largely predictable load of sewage and if there was to be any treatment it would only be sewage that was treated, not the stormwater also. The disadvantage was the extra costs involved in constructing two separate systems.

It had long been assumed that if sewage could be emptied into rivers it would soon be rendered harmless by the moving water, but such assumptions were challenged as ever-larger quantities were emptied into the same river at various points along its length by growing towns. Cities without a convenient ocean or river looked at ways of treating sewage. Chadwick himself was a great enthusiast for putting sewage to productive use in fertilising the land, thereby recouping some of the expenses incurred in the construction of sewers as well as the re-fertilisation of the soil which had been depleted by growing the food eaten by city dwellers. Of course, largely solid town wastes had long been used as manure in market gardens on the peripheries of towns: what was new and imperfectly understood was how to employ huge volumes of water-diluted sewage and the value it possessed as a manure (Goddard 1981: 32–6). Some towns in Britain did establish sewage farms and run them for a time but it was in Berlin and Paris and in Australian cities, notably Melbourne and Adelaide, that such farms were used extensively, in the way Chadwick had envisaged.

New knowledge of bacteria and, especially, of their role in breaking down and oxidising organic materials opened up new possibilities for sewage treatment. This led to the development of various methods of biological filtration in which sewage was sprayed over beds of stones, encouraging the growth of bacteria on the surfaces of the stones and the oxidisation of organic matter in the sewage. The activated-sludge process was an alternative and ultimately more-effective approach. It was developed from the observation that if some of the sludge that had settled out of a batch of sewage that had been aerated — that sludge consisting of active, living bacteria, hence the term 'activated sludge' — was introduced to a new batch of raw sewage, the process of aeration and stabilisation would proceed far more rapidly. This process was widely adopted from the 1920s in Britain and elsewhere. Leftover sludge was difficult to dispose of but it was discovered that if it was digested or fermented by anaerobic bacteria in a septic tank it was rendered inoffensive and could be used as a fertilizer. Karl Imhoff pioneered this process of sludge digestion on a commercial scale (Bruce 2 1968: 1387–94).

By the early years of the twentieth century, the inhabitants of cities in modern economies had largely overcome the challenges of living in close proximity to each other without succumbing to infectious diseases. There existed a set of interlocking technologies that could provide an ample supply of pure water for household use, for commerce and manufacturing and to fight fires. There were also effective ways of removing wastes along water-flushed sewers and of disposing of those wastes in ways that would not impact adversely on others. Cities knew about these advances, which were publicised in engineering journals and through the diaspora of increasingly well-trained engineers from England, Scotland and parts of Western Europe to countries of recent European settlement elsewhere in the world. This meant that it was possible to build water supply and sewerage infrastructure just about anywhere where materials, men, finance and an engineer could be brought together.

The distinctive pattern of technological development that led to this outcome constitutes an example of path dependence; that is, a situation where existing technologies shape and direct future developments, or, to put it in another way, technological change depends largely on its own past (Mokyr 1990:162–5). Once the Chadwickian system had demonstrated its superiority over earlier methods of water supply and waste removal, experimentation with alternatives to it virtually ceased. In the Netherlands in the late 1860s, Charles Liernur developed a pneumatic or suction system that was used on a significant scale in Amsterdam and in parts of Prague and St Petersburg but no other cities were persuaded to adopt it in preference to water-borne waste removal (Bruce 2 1968: 1384). Instead, innovators focused on the bottlenecks, the weak spots in the Chadwickian system in order to improve them; hence the developments sketched above. In this way the whole system became more tightly integrated and efficient.

This also meant that alternatives were less likely to be developed. Where a technology is expensive and disruptive to install, it is less costly to expand an existing system as the number of consumers grows rather than replace it with something different. Heavy investment in the internal combustion engine over many decades, for example, has discouraged large-scale investment in alternative forms of propulsion for cars such as electric or steam power. As an interlocking set of technologies develop around a particular way of, say, harnessing energy or removing domestic wastes, it tends to become self-reinforcing for other reasons also. New usage habits grow up around new technologies; for example, in this volume Graeme Davison discusses how people have consumed increasing amounts of water to satisfy a range of needs beyond what is required for drinking, cleanliness and waste disposal. Usage cultures then determine what is perceived to be an adequate supply. The creation of institutions that depend on the technology for their existence further reinforced commitment to the existing system; water supply and sewerage authorities, for example, were and are committed to sustaining the large centralised systems which they construct and manage.

Despite the dominance of the Chadwickian system, each city faced its own unique set of circumstances pertaining to climate, geology and location in relation to the availability of rivers and of rainfall. There were also varied governmental and financial constraints that shaped what was likely to happen. No cities adopted best practice along the whole range of technologies mentioned here, and some adopted very few. Our focus is on Australia and while I will use Melbourne as an Australian example and as the city I know most about, but it is only in the timing and detail of its development that it differs from other cities in Australia and elsewhere.

Initially, water was supplied to the rapidly growing settlement by water carts and from tanks filled from rainfall on roofs. As the city grew and polluted the Yarra River, from which water carriers drew their supplies, the system became increasingly inadequate. British migrants worried about the possibility of a cholera outbreak in Melbourne and by the later 1840s there was support for the construction of a city-wide system of supply to replace these private and decentralised approaches. Cost was a central issue and the debate was between a system that would pull water from the Yarra not far upstream from the city, but would require constant pumping, and one that would draw it from much further afield and would consequently cost far more to construct but would have much lower operating cost because it could rely on gravity, as the Romans had done in their pioneering water-supply systems. The latter was chosen and construction of the Yan Yean began against the backdrop of the Victorian gold rushes.

This activity benefited from the accumulation of medical and engineering skills taking place in England but also extended it in some ways. The English-trained chief engineer, 27-year-old Matthew Bullock Jackson, built an earthen dam at Yan Yean that was one of the longest then attempted, to create what was perhaps the world's largest reservoir at that time. He was elected as a member of the prestigious Institution of Civil Engineers in London for his work and for his paper describing the construction of Yan Yean. Jackson's brief stay in Melbourne was tumultuous, but it was for his achievement in building Yan Yean that he was included among water engineer and historian G. M. Binnie's handful of outstanding water engineers of the early Victorian era (Binnie 1981). Some locals worried that the dam wall was unsound but their concerns were unfounded and the embankment still impounds some of Melbourne's water. Local conditions were not well understood and some sceptics feared that during hot Australian summers the much higher rate of evaporation would result in a dry reservoir, but these fears too were unfounded.

Yan Yean massively increased the supply of water to Melbourne but it was not pure water. Lead from the reticulation pipes poisoned some Melburnians. More became ill over the next few decades because the water contained organic impurities. Only gradually, as knowledge of disease transmission became available and local investigations identified polluted catchments, could suspect supplies be eliminated and new clear water sources from catchments closed to all other uses be redirected into the Yan Yean reservoir. This became Melbourne's approach for the next century. It would harvest its water from closed catchments and so save the cost of water treatment. This involved intermittent fights with logging interests and irrigators, but the city succeeded in cornering enough pure water for itself until the 1970s (Dingle and Doyle 2003).

Melbourne's changing methods of waste disposal mirrored the experience of many other cities. Cess pits, located at the bottom of the garden where there was one, were widely used, as they had been in Britain. Poorly constructed and inadequately regulated, they were eventually banned in the city, though not in the suburbs. A system of pan collection replaced it. This worked effectively for some time but, by the 1880s, urban growth led to its breakdown, especially in the inner suburbs, as it became increasingly difficult to find locations where the growing tonnages of wastes could be deposited safely and effectively (Barrett 1971: 75–86, 127–137). Eventually it was agreed that something better was needed. A Royal Commission looked briefly at alternatives to water-borne systems and at criticisms of them, but there was overwhelming support for them (Third Progress Report 1889: xv–xvi). James Mansergh, an eminent British sanitary engineer asked by the Victorian Government to visit and recommend the most appropriate plan, explained that as 'nightsoil does not improve with keeping ... the true policy is to get it off the premises as rapidly as possible': Chadwick could not have put it better. Mansergh advocated a water-borne sewerage system. Local engineers had long asked for this too, but the Government listened to Mansergh (Report on the Sewerage and Sewage Disposal of ... Melbourne... 1889: 13).

There has been some debate as to why Melbourne was so slow in switching to a water-borne sewerage system. Was it an economically rational decision to change once the costs of the old system became greater than the new, as Gus Sinclair has argued (Sinclair 1975)? One critic has disputed that the evidence supports this conclusion (Merrett 1977). Or was it a fragmented structure of urban governance that delayed a decision until the threat was seen to be too great to be ignored any longer, as David Dunstan has argued (Dunstan 1984: 233–74)? The threat to public health was a factor, as it had been in England, although in Melbourne it was typhoid rather than cholera that was the scourge and there were epidemics in the 1880s which were not experienced by Sydney or Adelaide, which already had sewers. This was enough to persuade residents of the salubrious outer suburbs in Melbourne — who had long argued that they did not have the public health problems of inner areas and thus did not need to help finance a city-wide sewer network — that they were not immune from epidemics, especially as many of them travelled into the city to work.

The Melbourne and Metropolitan Board of Works (MMBW) was created, largely in the image of its London namesake, and began its work in 1891. A locally trained (Melbourne University) engineer, William Thwaites, was appointed chief engineer. While saying he was building Mansergh's plan, he went ahead and built his own, superior, separate system. The first connection was made at the All England Eleven Hotel in Port Melbourne on 17 August 1897 and rapidly thereafter most of the metropolitan area was connected. Perhaps the most interesting feature of Melbourne's scheme was its total reliance initially on a sewage farm to treat all wastes. This was adopted only because it would have been far more costly to have built a pipeline to the coast at a point where a significant tidal scour could have washed Melbourne's wastes out to sea, but it represented a serious attempt to realise Chadwick's vision of putting wastes to productive use, and its size attracted international visitors. Interestingly,

Melbourne built its sewerage system just before more-sophisticated methods of treatment had been developed. Had it begun two decades later it could have utilised an activated-sludge treatment plant from the outset (Dingle and Rasmussen 1991).

For the last century Melbourne has relied on expanding the systems it had in place by the beginning of the twentieth century to cope with its growth from a city of around half-a-million to one of three-and-a-half million at the beginning of the new millennium. For water supply it relied on building more dams. The engineers estimated future supply needs and planned and built the dams required to store the requisite amount of water to meet those needs. Yan Yean was augmented by storages at Maroondah and Silvan in the interwar years.

There was a major threat to the effectiveness of this approach after the Second World War. Rapid population and housing growth in the 1950s and 1960s, and a shortage of investment capital that had first been felt in the depression of the 1930s but was not eased until the 1970s, meant that storages and pipelines to bring the water to the city were both inadequate. The Upper Yarra dam was finally built in the 1950s but taps ran dry on hot days and frequently applied water restrictions forced people to use buckets instead of hoses to water their large gardens. Suburbanites were irritated by this but they were encouraged by the prospect of more dams and greater supplies in the future, as promised by the MMBW. The Greenvale and Cardinia dams were built after increasingly difficult political fights but other proposals were knocked back. The mighty Thomson Dam filled in the 1980s. It did indeed 'drought proof' the city until recently, as promised, but despite demand-management programs from the 1990s that have reduced the rate of increase of per-capita water use, Melbourne is once more on restrictions.

Melbourne's sewerage system was also expanded to meet growing demand whenever possible. The Werribee Sewerage farm proved capable of expansion for many decades; although the methods of treatment changed also as settlement lakes took over from land irrigation. As with water supply, the sternest test for the system came with the massive post-war expansion of the housing stock. Water supply was given precedence over sewerage so the new suburbs on the outer fringes of Sydney and Melbourne, especially, spread out with water but no sewers. The house blocks were large, though often not the quarter-acre block of legend. But they had room in the back yard for their own decentralised method of sewage disposal, the septic tank, as the planners intended. This represented a failure of the centralised system, not because of technical difficulties but rather because of the scarcity of investment funds. State governments and water-supply authorities threw the capital costs of what they believed to be an inferior system onto individual householders. The situation was not rectified until the Whitlam government pumped federal money into the eradication of the sewerage backlog in the 1970s. Subsequently, connections kept up with suburban expansion; indeed, houses could not be occupied until a connection had been made. A new treatment plant employing modern methods was built at Carrum in the 1970s as the suburbs continued to expand to the southeast.

The process of change in water supply and waste disposal for most cities, including Melbourne, has been from decentralised, often private and small-scale, provision to large-scale, centralised, publicly owned and highly capitalised infrastructure which, for the most part, has performed effectively for a century or more. This was part of a wider networking of nineteenth- and, especially, twentieth-century cities to provide public utilities such as gas and electricity, telephones, trams and trains (Tarr and Dupuy 1988). However, the future of centralised water supply and waste disposal now looks far less assured because it is difficult to see how they can be expanded further without exacerbating current problems. In common with other Australian cities, Melbourne is now chronically short of the quantities of water required to maintain present levels of consumption; hence continuing restrictions. With the prospect of more variable and lower rainfall ahead, the options for increasing supply have contracted. The scope for building new dams has virtually disappeared because viable sites have already been utilised and more harvesting from rivers will further degrade their flows, with adverse impacts on the downstream environment. The pipeline to bring water from north of the Great Dividing Range to augment Melbourne's supplies will make only a modest addition to requirements at the expense of alienating much of Victoria's non-metropolitan population, who see this as another attempt by Melbourne to ensure that it can continue to water its lawns. Even the currently popular desalination plant to be built near Wonthaggi will require massive energy inputs and consequent increases in greenhouse gas emissions. It is likely that any attempt to further increase water supply in the metropolitan-wide system — the traditional solution to shortages — will bring escalating environmental costs.

There have been modest attempts to manage demand and these have been modestly successful from the 1990s in slowing the growth in demand. These approaches appear to work best when people perceive that there is a crisis and they need to do something to lessen it. This is perhaps most dramatically illustrated in Brisbane's case, as discussed by Peter Spearritt in Chapter 2. Otherwise, the experts argue that while this is a worthwhile strategy it is unlikely to yield dramatic improvements (Troy, Randolph and Holloway 2007: 8).

The water closet and all that lies below it now also appears increasingly problematic. It is a major user of potable water in every household and large volumes of potable water need to be used and dirtied simply to keep the sewerage system flushing safely and effectively, as Troy has pointed out (Troy, Randolph and Holloway 2007: 9). In 2007, the ABC urged its listeners to attempt to make do with not more than 40 litres of water per person for two days. Many reported that this was not possible if the water closet was used to flush away faeces. It simply required too much water to operate effectively. At the other end of the line, the disposal of increasingly large volumes of Melbourne's partially treated effluent at the ocean outfall at Boags Rocks is having an adverse impact on the marine environment and there is growing opposition to its continued use.

Why then are we still using this technology that is so wasteful of an increasingly scarce resource? Path dependence does not guarantee the continuation of existing technologies if there are compelling alternatives. For example, steam as a source of power gave way to electricity and the internal combustion engine once the advantages of the latter were seen to be overwhelming. However, it would be difficult to argue that this is yet the situation with water supply and waste disposal. There is not yet the sense of complete breakdown that had persuaded cities in the past that the costs of continuing on in the same way are unacceptable and a new approach must be found, nor are replacement technologies clearly available. Do we then have to experience higher levels of dysfunction before there is sufficient impetus for change?

What we do have is a range of partial technologies that can economise on the use of potable water and the waste-disposal system. These range from harvesting rainwater from the roofs of homes for use on the garden, the toilet and the laundry to recycling greywater for similar uses. These have the capacity to significantly augment and extend the life of the existing centralised supply system. They are small-scale, decentralised and, at present, the costs of utilising them are met by households. Although governments have recently offered incentives for households to adopt these alternatives, they have not for the most part been espoused by the water-supply authorities or integrated into any overall strategy that can utilise both local and central technologies. This is the political challenge to change. With a co-ordinated effort by governments, public utilities and private suppliers it may be possible to avoid a costly major technological discontinuity and modify and add incrementally to what we already have. In this way the Chadwickian solution may yet enjoy an extended lease of life.

References

Binnie, G. M. 1981, *Early Victorian Water Engineers*, Thomas Telford, London.

Bruce, F. E. 1, 1968, 'Water Supply', in C. Singer *et al.* (eds), *A History of Technology*, vol. 5, Clarendon Press, Oxford.

Bruce, F. E. 2, 1968, 'Water Supply and Waste Disposal', in Singer *et al.* op. cit., vol. 4.

Chadwick, Edwin 1965, 'Report on the Sanitary Condition of the Labouring Population of Gt. Britain', Edinburgh University Press, Edinburgh.

Derry, D. K. and Williams, Trevor I. 1960, *A Short History of Technology from earliest times to A.D.1900*, OUP, Oxford.

Dingle, Tony and Doyle, Helen 2003, *Yan Yean: A history of Melbourne's early water supply*, Public Record Office Victoria, Melbourne.

Dingle, Tony and Rasmussen, Carolyn 1991, *Vital Connections: Melbourne and its Board of Works 1891–1991*, McPhee Gribble, Melbourne.

Dunstan, David 1984, *Governing the Metropolis. Politics, Technology and Social Change in a Victorian City: Melbourne 1850–1891*, MUP, Melbourne.

Goddard, Nicholas 1981, 'Nineteenth Century Recycling: The Victorians and the Agricultural Utilisation of Sewage', *History Today*, June.

Lampard, E. 1973, 'The Urbanizing World', in Dyos, H. J., and Wolff M. (eds), *The Victorian City: Images and Realities*, vol. 1, Routledge and Kegan Paul, London.

Merrett, D. T. 1977, 'Economic Growth and Well-Being: Melbourne 1870–1914: a Comment', *Economic Record*, vol. 53.

Rosen, G. 1973, 'Disease, Debility and Death', in Dyos and Wolff *op. cit.* vol. 2.

Tarr, Joel A. 1988, 'Sewerage and the Development of the Networked City in the United States, 1850–1930', in Tarr, Joel A. and Dupuy, Gabriel (eds), *Technology and the Rise of the Networked City in Europe and America*, Temple University Press, Philadelphia.

Mokyr, Joel 1990, *The Lever of Riches: Technological Creativity and Economic Progress*, OUP, New York.

'Report on the Sewerage and Sewage Disposal of the proposed Melbourne Metropolitan District' (1890), Victorian Parliamentary Papers, vol. 2, Melbourne.

Sinclair, W. A. 1975, 'Economic Growth and Well-Being: Melbourne 1870–1914', *Economic Record*, vol. 51.

'Third Progress Report of the Royal Commission to Inquire into and report upon the Sanitary Condition of Melbourne', 1889, Victorian Parliamentary Papers, vol. 4, Melbourne.

Troy, Patrick, Randolph, Bill, and Holloway, Darren 2007, 'A New Approach to Sydney's Domestic Water Supply problems' typescript.

Wohl, Anthony S. 1983, *Endangered lives: Public Health in Victorian Britain*, Dent and Son, Guildford.

Chapter 2

The water crisis in Southeast Queensland: How desalination turned the region into carbon emission heaven

Peter Spearritt

If you flew into Brisbane in 2006 you were greeted with a huge advertising hoarding on the airport drive with the words:

'Head to Queensland. The climate's great for growth.'

The text was set to a backdrop of vast humanoid cranes walking across a brown landscape. The state government advertising campaign, run nationally, reminded punters of Queensland's booming open-cut coal mines, and set the scene for a state on which the sun never sets. Demographers, senior public servants, property developers and Premier Beattie bragged that SEQ was the fastest growing area in Australia, with more than 1000 people a week moving north. Local morale was further boosted by Sydney's ailing property market and Melbourne's unpredictable weather.

Such optimism has long marked image-making in Queensland, ever since the northern state stole the surfing limelight from New South Wales in the 1950s and 1960s. With the spectacular growth in tourist numbers in the last 40 years, Queensland put great store in its advertising slogans: 'Beautiful one day, perfect the next' is now one of the world's most long-lived tourist marketing ploys. Coastal Queensland houses over one-third of all holiday-let rental apartments in Australia.

So in Queensland, the climate is good for holidaymakers, for locals and 'for growth', the optimum trilogy. What could go wrong? No-one ever suggested at the countless focus groups run by political parties that the cities might run out of potable water. Yet that is precisely the scenario that confronted Beattie in his last year of office, and has had water authorities, politicians and senior bureaucrats ducking for cover ever since. The ad agency that thought up the dusty imagery for 'The climate's great for growth' must have known something.

While all major Australian cities have experienced periods of drought, few have been more innocent of the possibility in the last few decades than Brisbane. The 1974 flood, which entered much of the CBD and all of the upriver suburbs and catchments, reinforced the notion that oversupply was the problem, not a lack of water. Over 90 per cent of the metropolitan area didn't get water meters until the 1990s — so the majority of Brisbane City Council (BCC) ratepayers simply paid an access charge for their water, not a usage charge. They had no way of knowing how much water they were using or wasting. Today, the hastily constituted Queensland Water Commission (June 2006) hands out egg-timers with a suction cap to be affixed to your bathroom wall to encourage you to shower in under four minutes. Such tactics have greatly reduced demand, and are being followed with interest around the world.

The turnaround in Southeast Queensland has been extraordinary, which is why it is such a pertinent study at present. Brisbane, from the 1960s to the 1970s, went from a city where most new subdivisions were unsewered to a sewered metropolis, the ALP City Council getting a large helping hand from the Whitlam government. But, at the same time, Brisbane had more fire hydrants than water meters. In the late 1980s, fewer than one in 15 houses had a water meter. No wonder per-capita consumption per day hovered around 700 litres. By 2007 — with restrictions and a well-supported public campaign — consumption had fallen to 140 litres, one of the lowest in the developed world, lower even than Israel. (See Cole 1984; *Courier Mail* 28 August 2007.)

Turn on the sprinklers, bury the water meters

In the 1950s, 1960s and 1970s, Brisbane's water debates were dominated by occasional bans on sprinklers, the adequacy of the system of reservoirs to supply sprinkler demand, an ongoing debate about sewering more houses (most post-1950 subdivisions were septic) and from 1974, the fear of flood, after Brisbane experienced its worst flood since the 1890s.

The proportion of properties with water meters fell from 80 per cent in the late 1930s to 6 per cent in the 1980s. Installing meters to the hundreds of thousands of un-metered properties became such a political hot potato that both sides in Council promised not to install more meters. In late 1984 thousands of brand-new water meters were buried at Boondall, a tip and wetlands area north-west of the airport. They await liberation by future archaeologists. Neither the dominant ALP group in Brisbane City Council (Australia's only metropolitan-wide council) nor its opponents wanted to be seen to be measuring, let along charging individual ratepayers for the amount of water they used. Water — as much of it as any household or business wanted or needed — was seen as an inalienable right.

The policy of not bothering to install water meters, even for new subdivisions, continued until the late 1980s. This changed in 1989, and in the six years between 1990 and 1995 218 000 new water meters, in black plastic containers flush with the nature strip, were installed throughout Brisbane. Suddenly households could be informed of how much water they were actually using. But this didn't stop the sprinklers, because water was plentiful and extraordinarily cheap, less than $1 for 1000 litres. The catchment areas and the dams, which filled from time to time with deluge rains, seemed able to keep up with demand (BCC Archives; *Brisbane Statistics*; *Brisbane Yearbooks*).

Water restrictions: How strict can you get?

For a society which just three years ago turned on water sprinklers at any hour of the day or night, and which still regards the installation of a swimming pool as the ultimate expression of sustainable sub-tropical living, Level 3 water restrictions, introduced in June 2006, banning any form of hosing at all, were an almost unthinkable nightmare. Earlier residents of Brisbane got stuck into the rainforest, harvested the fine timbers, and built sensible wooden houses until generations of sawmillers ran out of trees to cut down. The imported species that give Brisbane what is left of its green canopy — especially the Leopard trees — struggled in the drought. Even the fig trees were troubled (Spearritt 2003). The city's fountains remain waterless, in a rather pathetic attempt to pretend that everyone is tackling the crisis seriously, a bit like the marketing ploy of buying carbon offsets to discount guilt when travelling by air.

Gone are Brisbane's long-touted ambitions as Australia's only sub-tropical capital city. And on the Gold and Sunshine Coasts the outdoor showers — surely the quintessential Queensland act for locals, interstate and international visitors — were turned off. To get an outdoor shower you had to cross the border and head for Byron Bay, where it still rains regularly and where much more of the natural landscape has been preserved in national parks and nature reserves. Byron Bay could easily have become another Gold Coast were it not for interventions from both a brave local council and the NSW state government, which under Premier Carr, attempted to prevent coastal overdevelopment.

Catchments for real estate, not for water

How did the SEQ crisis come to pass? Why did a government whose own Population and Forecasting Unit correctly predicted the rate of growth in Southeast Queensland allow water stocks in its dams to fall to perilously low levels while simultaneously allowing the government's own electric power stations to continue to thrive on potable water supplies? By 2006 the power stations were using up to one-fifth of SEQ's daily consumption of water. What an irony that at peak times in summer most of the electricity goes to air-conditioning for apartments and houses built without any attempt at cross ventilation. Surely the upper-middle class should be left to sizzle in the neo-Tuscan mansions they erect, having managed to get rid of the traditional Queenslanders on river and hillside sites. And investors in inner-city apartments should be made to live in their shoddily-designed concrete boxes after the air-conditioning has been disconnected.

The explanation of the crisis in SEQ lies in an analysis of infrastructure quick-fixes popular with civil engineers, a remarkable lack of accountability in the water bureaucracies, and untold arrogance in the electricity authorities. Local governments and the state government were so preoccupied by real-estate-driven growth that they lost interest in the quality of the urban environment. The property industry in Queensland is the largest single source of election funds for both the ALP and the coalition parties. Since the late 1950s all Queensland premiers have embraced the property industry as a vital engine for growth.

Premier Bjelke Petersen fondly bragged about the cranes on the skyline. Subsequent premiers have been more subtle but the message has been the same. Premier Beattie finally delivered the first statutory plan for the SEQ region in June 2005, partly to placate growing environmental concerns about the fate of the coastal and hinterland landscape, but the real beneficiary has been the property industry, which now has oodles of infrastructure support to create higher densities in the inner suburbs and certainty about where it can bulldoze afresh to create instant 'Lakes estates' on the urban periphery. Fortunately some of the lakes do offer on-site stormwater management and capture.

A society so dependent on real and speculative building booms creates a haphazard urban form. Car-based suburban development, well beyond any prospect of either rail or Brisbane's impressive and recently augmented busway system, has led to the creation of a 200 kilometre city from Noosa to the Tweed. As over 30 per cent of the population can't drive or don't have a car, there are a lot of youth and older adults stranded in this urban form. Successive state governments refused to contemplate the looming problem and no major political party has yet been brave enough to question the rights of an adult electorate where three-quarters of voters drive (Spearritt 2004).

Only 17 per cent of Southeast Queensland is held in state forests and national parks, compared to 43 per cent of Greater Sydney. One obvious result is that the catchment areas for dams in SEQ are not a patch on the Sydney catchment areas. Because so much of the relatively arable environment of SEQ had been carved up into small rural landholdings by the early 1950s, when it came to locating new dams they ended up to the north-west of the city in a relatively dry catchment area, the Wivenhoe dam site selected as much to prevent flooding as to collect and store water. Add the worst drought in 100 years, and you've got a very big problem (Brisbane Institute 2003).

Existing bulk water supply and transport infrastructure, *SEQ Draft Water Strategy***, April 2008, chapter 5, p.105**

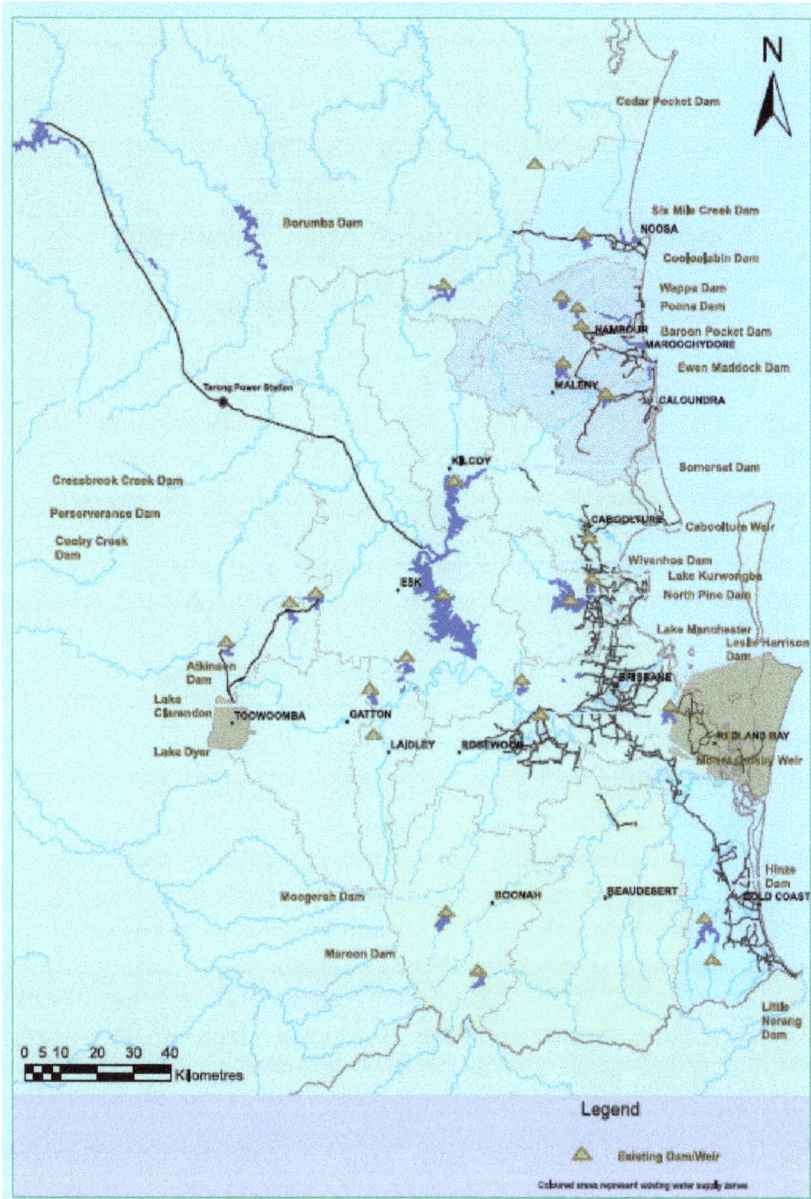

Even Gold Coast property developers got worried at the thought that the water might run out. Imagine the indignity of having to buy in water — via truck — from northern New South Wales to fill up your lap pool. It hardly goes with the Gold Coast's image of sunshine, instant palm plantings and unlimited largesse, from meter maids and schoolies week to champagne at the Indy 500. The current Gold Coast marketing campaign, 'Very GC', brags of 'miles of sandy beach, lush green rainforest, world-class golfing greens and world-famous theme parks'. (Verygc.com 2007)

Whatever happened to the 'deluge rains'? Belated recognition of the water crisis

The Beattie government got rattled by the water crisis. The government had already weathered the doctor shortage (and embarrassment about some surgeons with below-par success rates) but managed to avoid too much blame for alleged shortcomings in electricity infrastructure, primarily caused by a rapid rise in air-conditioning demand (AJPH, 2004+). But unlike these two issues, every member of the public knew about the water crisis, for the remarkably obvious reason that we were not getting much rain and, in particular, were not getting the 'deluge rains' which, every few summers, used to augment the Wivenhoe and Somerset dams and freshen up their gardens. Suburban streets the length and breadth of Australia's 'fastest-growing urban region', as the Beattie government was wont to boast, rang out with neighbourly exhortations for rain. Nature strips, once watered, were now brown, and so were the lawns. Hardy shrubs gave up. Nurseries closed, car-washes flourished (now they have to use recycled water), swimming-pool builders grumbled and landscape gardeners struggled to make ends meet.

Successive senior bureaucrats, government ministers and their advisors ignored the warning signs, sounded as early as 1997 by experts in the Department for Natural Resources and Mines, which in various guises had the biggest group of hydrologists and others responsible for assessing water resources and calibrating those resources with consumption patterns and evaporation. The failure to follow this advice reflects badly on senior government bureaucrats and equally badly on ministers who encouraged a culture of 'see no problems, speak no problems' (Interviews 2007–08).

The Department of Natural Resources released a draft strategy for water supply in SEQ in August 2004 and a much more alarmist, but well-argued, Interim Report in November 2005, which included the — at that time — amazing proposition that consumption might have to be limited to 300 litres per person per day. As the Executive Summary put it: 'If significant inflows to the Wivenhoe, Somerset and North Pine are not received by around February 2006, SEQ will be in the grip of the worst drought in recorded history.' The report pointed out that these dams were last full in February 2001 and had only minimum inflow in 2004. By November 2005 the dams were under 35 per cent full. The report's main recommendations included: water restrictions, recycled water for industry and power stations (to reduce demand on Somerset and Wivenhoe), construction of a weir on the Mary River, and the investigation of 'regional desalination facilities' (SEQRWWS *Interim Report*, November 2005: 1).

In the following months, dam levels continued to fall precipitously. Generous state government and local council tank subsidies were introduced, enabling householders who installed more than 5000-litre capacity to recoup up to $2200. The Beattie Government ran full-page newspaper advertisements exhorting residents to install tanks and take advantage of a subsidised scheme to install water-wise devices. Tens of thousands of households took up the offer. All new government, commercial and residential structures were encouraged to collect rainwater on site. Level 2 water restrictions, which had been introduced in October 2005, were made more stringent, with Level 3 introduced in June 2006 (hoses banned) and Level 4 in November 2006, allowing bucket-watering for just a few hours a week. Greywater recycling for gardening purposes became legal under BCC regulations in late 2006. Brisbane, a dusty city, especially in dry winters, became dustier still as the brave new world of freeway tunnelling projects, proclaimed by Lord Mayor Campbell Newman, with the implicit backing of the State Government, created huge piles of shale and dirt. Cynics wondered out loud who would tunnel for cars at a time when the very supply of adequate water for the metropolis hung in the balance (Pretty 2006; Dixon 2005).

'Poowoomba' and the politics of recycling

The garden city of Toowoomba — famous for its floral festivities — was in even more trouble than Brisbane. Its citizens had already been warned by the Department of Natural Resources and Mines in December 2004 that its extraction procedures for underground water were exceeding safe water yields. The Toowoomba City Council sought support from the federal government's new National Water Commission for a water recycling grant, supported by Minister Malcolm Turnbull on the proviso that they held a referendum.

Local business interests, led by millionaire property developer Clive Berghoffer, a former National Party MP and medical philanthropist, organised 10 000 signatures from 'Citizens against Drinking Sewage'. This clever if misleading notion, following on the success of 'Australians for a constitutional monarchy' (we don't want a 'politicians' republic') underpinned the successful 'No' campaign in the referendum. A 6–3 vote at the Toowoomba Council in favour of recycling and a vast scientific effort mounted in favour of the 'Yes' vote proved to be of no avail. Beattie did not help the situation by quipping on 2 June 2006, probably inadvertently, that 'replenishing dams with recycled water' would be 'the Armageddon solution'. Media-savvy Beattie always had a penchant for a slogan but this one backfired (*Courier Mail*, 3 June 2006).

The 'Yes' case rested primarily on returning recycled water to dams, multiple and proven barrier-treatment processes, with two–three years' testing by CSIRO before release. The 'Yes' case, championed by Mayor Di Thorley, drew modest support from Beattie, Turnbull and the *Courier Mail*. The 'No' case — 'to deny your natural instincts and adopt untested new technology is foolish' — went on to claim that thalidomide, asbestos and mad-cow disease were all caused by the ignorance of 'the long-term effects of science'. On 29 July 2006, 62 per cent of the population voted against the referendum question 'Do you support the addition of purified recycled water to Toowoomba's water supply'. Toowoomba, with its population of 120 000, remains desperately short of water (Vuuren 2007).

Beattie's water grid to the rescue

In August 2006 a re-branded department, now the Department of Natural Resources and Water, issued its *Water for Queensland, a long term solution* and gave Beattie the concept of the 'water grid' as the way forward. The analogy with the electricity grid amused some commentators, even though electricity is rather easier to manufacture than potable water and comes with a vast distribution network already in place. Imagine the Government's embarrassment when, at the budget estimates committee, it was forced to admit that its own Tarong Energy Power Station had been secretly 'siphoning' potable water from Wivenhoe despite an edict from Energy Minister John Mickel that it should take water from Boondooma Weir. The press leaped on the revelation, so an 'agreed separation' was promptly organised for Tarong CEO Andrew Pickford. Power station operators prefer potable water for their cooling towers because it is less salty than other water sources. But this revelation came at a time when all of Southeast Queensland was being asked to show restraint while government-owned instrumentalities obviously went their merry way (*Gold Coast Bulletin*, 15 July 2006; *Courier Mail*, 20 July 2006; *The Australian*, 20 July 2006).

The tone of *Water for Queensland* was grim: 'If Queenslanders are to maintain the lifestyle they currently take for granted, it is essential that demand for water is reduced and supplies are increased, so that economic growth and wealth creation can continue.' This document may well constitute the emergence of the ALP as a 'lifestyle' political party. The document boldly acknowledged the defeat of the Toowoomba referendum but stated that 'recycling within residential and non-residential developments will need to be introduced'. It gave Beattie the 'water grid' terminology, the catchphrase to solve everything, and announced the following infrastructure for 'the short term to 2016': Gold Coast desalination facility (45 000 ML/a); Western Corridor recycling scheme (30,000 MLa); Traveston dam stage I (70 000 ML/a); and a raft of smaller projects.

The Government paid lip service to some of the environmental impacts, especially of the desal plant, where the contemplation of alternatives got short shrift, while the Traveston dam proposal attracted a voluminous report with dozens of mitigation measures. The report pointed out that only 6–7 per cent of treated effluent in SEQ was currently recycled, mainly for golf courses and sports ovals. It also pointed out that the Western Corridor recycling scheme would make water available to the Tarong, Tarong North and Swanbank power stations. It had much less to say about how much electricity would be required to move all this water around SEQ, including the fact that the pipes will need continuous water flow to remain operational (*Water for Queensland* 2006: 1, 31).

Beattie's 'water grid', and an extensive advertising campaign for water-wise initiatives and tank subsidies, got him off the water hook for the September 2006 election and the ALP won a fourth term with little loss of seats. The National Party had failed dismally to command attention on the water issue, shooting itself in the foot when one of its senior politicians, Lawrence Springborg, suggested that evidence that male Danish fish developed female characteristics when swimming in recycled water could have implications for 'feminisation' in Queensland. The junior party in the coalition, a rag-tag of Liberal members, simply couldn't get their minds around either the scale or the severity of the water issue (*The Australian*, 1 August 2006; *Courier Mail*, 21 November 2006).

Despite the disarray of the opposition, pressure on the government continued to mount, not least because the dam levels kept falling, unlike the rain. In late October, less than two months after the election, Beattie announced that he would hold a referendum on recycling in the coming year. The Southeast Council of Mayors — nervous after the Toowoomba result — said they would not take sides in the referendum, though Brisbane Lord Mayor Campbell Newman came out in favour of recycling. Beattie abandoned the referendum idea in late January, explaining that the situation was so dire that purified recycled water 'is no longer an option, we have no choice'. He also explained that the Queensland Water Commission had given him 'compelling advice' to cancel the 17 March plebiscite. For once, Beattie got a favourable editorial in the *Courier Mail*: 'With Brisbane's Wivenhoe Dam at just above 20 per cent capacity, Premier Peter Beattie has made the right decision to press ahead with recycled drinking water for southeast Queensland and scrap what would have been a farcical $10 million plebiscite over the issue.' Four days later the *Courier Mail* informed its readers that some of them were already drinking recycled water (Beattie press release, 28 January 2007; *Courier Mail*, 28 October 2006, 28 December 2006, 29 January 2007, 3 February 2007).

A handful of well-informed journalists continued to point to the failure of the state government's water policy initiatives much more effectively than the divided and demoralised opposition. In April 2007, Craig Johnstone told his readers:

Remember how we were told nine months ago that drinking recycled waste water was an Armageddon solution? Next year, we'll be puckering up to glasses full of it. Remember last year, when the Government insisted its future water supply planning was sound because it was based on 2004–05 inflow figures into the dams, which were the lowest on record? It turns out the 2006–07 inflows are half that figure. Policy options that were beyond the pale six months ago are suddenly central to drought-proofing the region.

(*Courier Mail*, 11 April 2007)

The Queensland Water Commission

The Queensland Water Commission was created by Beattie on 19 June 2006, just before the Toowoomba referendum. Chaired by Elizabeth Nosworthy, a well-regarded and no-nonsense corporate bureaucrat, it was primarily set up to dictate uniform water restrictions in Southeast Queensland and to oversee the claims and activities of the various water authorities, which Malcolm Turnbull, among others, had criticised for self-interest and income maximisation at the expense of sound water policy.

The Southeast council mayors had been bickering for months about uniform water restrictions and they would often break ranks. Outrage greeted Gold Coast Mayor Ron Clarke when he allowed his own residents a 'wet weekend' of hosing down their driveways and washing their cars in May 2006, simply because the Hinze Dam happened to be full. But Clarke's action reflected the local view voiced by many Gold Coast residents that they should not be dictated to by Brisbane. The complications of overlapping jurisdictions and financial responsibilities in the water bureaucracies are much less tractable. The Queensland Government is now proposing to buy out the interests of local governments in water, but councils are bitterly complaining that the remuneration is insufficient for both the asset value and its long-term income potential. Owning water and charging for it, along with sewerage provision, has formed a major part of councils' urban assets and cash flow and has been a central plank in the services they provide for charging rates (*Courier Mail*, 21 April 2006. Between 1 May and 25 May 2006 the *Gold Coast Bulletin* continually reported the water-restriction issue).

The phenomenal success of the Queensland Water Commission in its 'Target 140' campaign, with Brisbane now boasting the lowest per-capita use of any major Australian urban area, shows just how much consumption can be reduced with media support and a degree of bi-partisan consensus, in marked contrast to the recycling referendum in Toowoomba, where cashed-up opponents ran a brilliant negative campaign as outlined, implying to the populace that they would literally be drinking from their own toilets. The negative campaign was aided and abetted — perhaps unknowingly — by sub-editors and picture editors who simply couldn't resist pictures of toilet bowls linked to taps. This facile visual journalism cut across the serious discussion of the water crisis to be found in the same papers.

The Queensland Water Commission continues to run a clever and successful marketing campaign for its Target 140. The Commission's website carries weekly updates on dam levels and household consumption. One of its key selling points is that if residents can stick to 140 litres, they will still be allowed to water plants with a bucket or watering-can for a few hours a week. This approach has the great merit that the Commission, while attempting to severely curtail household usage, is not trying to mandate exactly how and where individuals can use what is deemed as a reasonable, if heavily constrained, daily rate of consumption. There is an element of trust in this that has paid off in a level of region-wide compliance that is quite remarkable. Council mayors are mightily relieved that the Commission takes responsibility for determining water restrictions. Gold Coast Mayor Ron Clarke wanted to break ranks again in January 2008 when heavy summer rain overfilled the (small) Hinze Dam and he suggested that allowing ratepayers to hose down their driveways would reduce the risk of flooding. The Water Commission did make one minor and sensible concession: Gold Coast beaches could again turn on their outdoor showers (*Gold Coast Bulletin*; *Courier Mail*, January 2008).

The Gold Coast desal plant

In 1994 the Albert Shire Council and the Gold Coast City Council (they were amalgamated the following year) produced a 14-page glossy brochure entitled *Water...Lifeline of a City*, replete with photographs of the Hinze Dam wall, a natural waterfall, sprinklers on golf courses, and the obligatory swim-suited woman lolling on a red flotation device in a pool. As with most such brochures, there was a brief explanation of the water cycle and of Australia as the driest inhabited continent, before readers were informed that a reliable water supply is 'vital' for 'the nation's most popular holiday region'. The little Nerang Dam and the augmentable Hinze Dam (about to have its wall raised for the third time) were said to provide for the needs 'well into the next century'.

The brochure also devoted quite a few pages to saving water. With 60 per cent of water used inside their homes, householders were advised not to use their toilet bowl as a bin or ashtray; take shorter showers; and check their taps for leaks. The 40 per cent of water used on the garden could be reduced by soaking, not spraying; using mulch; adding a timer to the sprinkler system; letting the lawn go brown in summer; and installing a swimming pool cover. They were even shown how to read their water meter.

(L) Front cover of *Water...Lifeline of a City*, **Albert Shire Council and Gold Coast City Council, 1994**

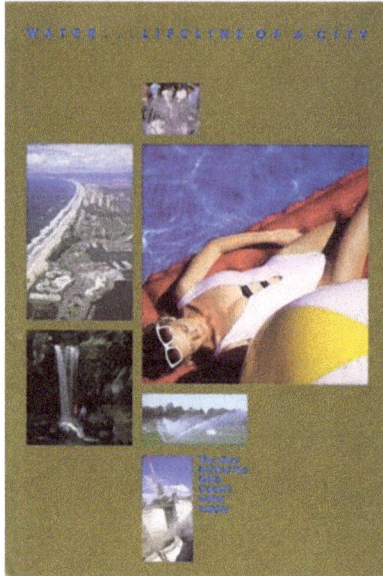

(R) Hinze Dam propaganda, from *Water...Lifeline of A City*, **Albert Shire Council and Gold Coast City Council, 1994**

Because the Gold Coast has a higher average rainfall than Brisbane, the Hinze Dam fills quickly, but because it is a small dam it also empties quickly. The Gold Coast continued to draw water from Wivenhoe, but once levels fell below 30 per cent the Gold Coast looked like it would be hung out to dry, to use a technical engineering term. The Goss Labor government, having abandoned the proposal to build the Wolfendene dam in the early 1990s, had not left the Gold Coast with a conventional legacy of large urban dams. With a population of 500 000, and more real-estate spruikers per head than anywhere else in Australia, investors got worried, as did their backers, banks and the superannuation funds.

By early 2005 the Gold Coast was well on the way to the strictest water restrictions in its history. Lord Mayor Ron Clarke announced in April that Southeast Queensland needed at least six mini-desalination plants, but a Brisbane City Council spokesman pointed out that using recycled water at Swanbank power station would save as much potable water as one of Clarke's projected plants. By September 2005 Clarke was pushing heavily for a fast-tracked desal plant, allegedly necessary because the coast's population would increase from 500 000 to 1.2 million within 50 years. The Gold Coast Council decided to bankroll a $165 million desalination plant to create a 'bulk water source', 'regardless of the drought'. As part of the water grid, the state government agreed in June 2006 to partner with the Gold Coast City Council and in November that year they formed a 50–50 joint venture company to develop and own the desalination plant, to be built on council land to the immediate west of the Gold Coast airport. This plant would have a capacity of 125MLs per day. The promotional video for the site, with its intake off Tugun beach, describes the project as 'environmentally sound and sustainable' while admitting that the desal water will be 'so free of salt and minerals' that 'minerals will have to be added' for potable consumption (*Courier Mail*, 27 April 2005; *Gold Coast Bulletin*, 24 September 2005; 26 November 2005. See also www.desalinfo.com.au, including the project's *Community Newsletter 2007*).

A bold and well-prepared state government and a similarly well-informed Gold Coast City Council could have implemented a serious water-tank initiative three years ago, retaining reticulated potable water supplies for kitchen and bathroom. The 150 000 dwellings on the Gold Coast that could be fitted with 20 000 litre capacity (gardens to be watered from laundry greywater) could have been undertaken for a maximum cost of $750 million over that period, including the cost of a pump and plumbing in for toilet and laundry use. This is calculated on a generous basis, with mass-purchase discounts of $5000 per dwelling. Instead, we get the desal plant at $1.2 billion, not counting the operating costs, let alone the carbon emissions. The Gold Coast desal plant is a knee-jerk instant fix. It proceeded without any environmental impact statement. Beattie told one protestor: 'If we don't have desal, we're not going to have any water. If you don't have water, you're dead.' Such insights appear to be propelling Labor premiers everywhere to embrace desal, which has almost become a plank of ALP platforms. (See Cooley *et al.* 2006; *Courier Mail*, 1 February 2007; Warren 2007.)

A really substantial water-tank initiative would have the added advantage that the thousands of sub-contractors on the Gold Coast currently employed in installing swimming pools, spa baths and Grecian bathrooms could be doing something environmentally useful. If you think I'm exaggerating you might like to contemplate Jade, a brand-new, one-apartment-per-floor block, near Q1 which is the world's tallest residential tower, explicitly planned as such, with a commercial observation deck. In Jade, which is right on the beachfront, every individual apartment has its own lap pool, saving its occupants the 30 seconds it would take to walk to the surf. One has to wonder about any society that embraces such conspicuous, privatised opulence, beyond anything imagined by the Romans in their baths.

The army of consultants, including many our of leading engineering firms, hired to justify the new infrastructure developments in SEQ specialise in going into enormous and lucrative detail about mitigating environmental impacts, and dismiss in just a couple of pages the prospect of much more extensive use of water tanks. Two pages is all that Sinclair Knight Merz devote to water tanks in their 1600-page justification of the Traveston Crossing Dam on the Mary River near Gympie. The consultants point to energy costs for pumping tank water, but don't compare that to costs for pumping the reticulated and recycled supply and fail to point out that gravity feed will be sufficient for some household tanks, depending on their location and the lay of the land. They are concerned about tanks getting contaminated, but there is no mention of the filtering systems now readily available. They are particularly concerned about tanks in a 'reduced rainfall scenario', an absurd comment when one reflects that the Sunshine Coast, Brisbane and the Gold Coast all have much higher rainfalls than the Wivenhoe catchment (Sinclair, Knight, Merz 2007).

Perth's desal plant opened in the southern suburb of Kwinana in November 2006. The WA Premier proudly proclaimed that Perth, in 'harnessing water from the ocean', had acquired 'an abundant source of drinking water that is not dependent on rainfall'. Although it will supply 17 per cent of the city's needs, if potable water in Perth were used only for the kitchen and bathroom it would not be needed at all. Perth's residents had been warned by George Seddon in his 1970 book *Swan River Landscapes* that they needed to 'fear the hose' and create gardens suitable for the landscape and the climate. The WA government cleverly side-stepped carbon emission criticisms by drawing electricity for the plant from the Emu Downs wind farm 200 kilometres north of Perth. This is sheer sophistry, as the power generated could equally be used for other needs. (For Perth's plant see www.watercorporation.com.au)

Resorting to desalination plants constitutes one of the great public policy failures of our times. Labor governments in Queensland, NSW, Victoria and WA, increasingly keen to prove how pro-business they are by placating their property industry lobbies, have gone down the track of desal plants with remarkably little analysis about the longer-term implications for both demand management and environmental costs. The freeway systems of the 1960s to the 1990s received much more internal government and public scrutiny than the desal plants have. One well-placed energy analyst has calculated that the proposed Sydney desal plant is the equivalent of adding 220 000 cars to Sydney's roads each year (Australia Institute 2005).

In embracing its desal plant, the Gold Coast City Council can now proudly claim to be Australia's least sustainable major city. With less than 2 per cent of its travel by public transport, its heavy reliance on air-conditioning and its desal plant, residents of the Gold Coast will shortly produce more carbon emissions per head than any other major Australian city. What a great claim for Australia's surfing holiday capital. How will the spin doctors respond? Perhaps their next advertising campaign will be about the Gold Coast being less environmentally conscious than Dubai.

Potential desalination sites, *SEQ Draft Water Strategy*, **April 2008, chapter 5, p.125**

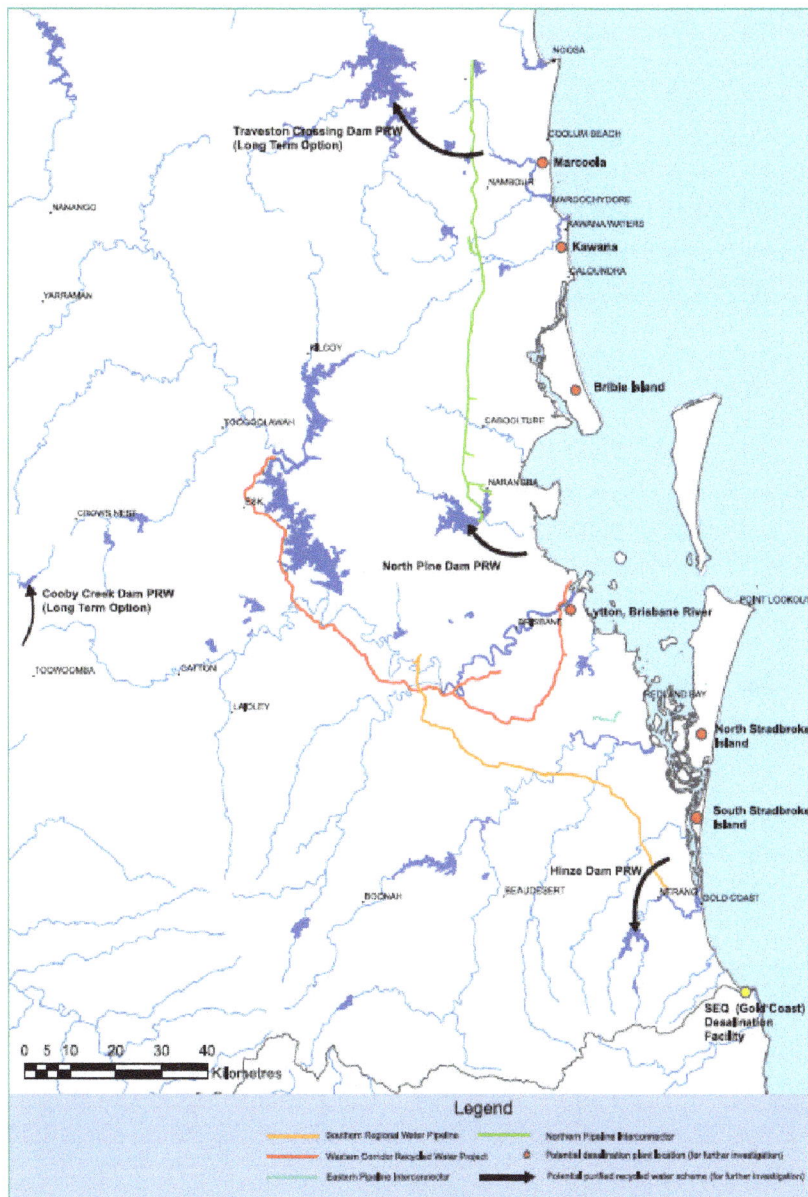

All this is the more extraordinary because both the Sunshine Coast and the Gold Coast have regular and quite healthy rainfalls, with the potential for householders to capture rainwater and for councils to harvest at least some stormwater. The Gold Coast and the Sunshine Coast received so much rain in January and February 2008 that every tank could have been filled and refilled within days. Such are the ironies of knee-jerk and alarmist infrastructure developments which do not adequately address alternative options. In going down the high-capital, high-energy and high-carbon emissions road, Queensland has now committed a generation to paying through the nose for a desal quick-fix rather than confronting more-sustainable approaches to climate change.

Conclusion

It took a long time for politicians, senior public servants and rarely-accountable heads of key agencies — from electricity power stations to water bureaucracies — to face up to the severity of the crisis in Southeast Queensland. For a while it looked like they would contemplate some sustainable initiatives, including what appeared to be genuine encouragement for the mass installation of water tanks. Had they been quicker to realise the severity of the issue, it would have been more sensible to suggest that households install 10 000 to 20 000 litres to get a rebate, and it is not too late for policy change to happen,

Instead, the Queensland Water Commission, in its latest Draft *SEQ Water Strategy*, released in April 2007, has in effect given up on rainwater tanks by suggesting that they can only provide 7 per cent of the region's needs by 2256. The Commission suggests, instead, another six desal plants up and down the coast, all located in the region's diminishing open space, including two of its sand islands. Yet most of coastal SEQ has, and is predicted to continue to have, a reasonable rainfall (The Draft Strategy is available at www.qwc.qld.gov.au).

The documentation on this public-policy failure is primarily to be found on the web in endless numbers of documents carefully re-phrased by wordsmiths to avoid later attribution or retribution. Senior public servants have become remarkably risk-averse in a political climate where governments hide behind FOI legislation and where speaking your mind is not encouraged. A number of university water-research centres get funding from firms who stand to make millions of dollars out of desalination infrastructure.

In this melange of public-policy obfuscation, a few brave acts have happened. Beattie cleverly abandoned his proposed referendum on recycling in January 2007 and the hapless Liberal/National Party coalition was unable to turn it into a telling political issue. Much less edifying is the rush to build desal plants with little attempt to inform the public about the high carbon emissions and high running costs of such installations. I wonder whether the Gold Coast desal plant, tucked away at the back of the Coolangatta airport, will be quietly decommissioned within a decade on both environmental and cost grounds, assuming an aeroplane overshooting the runway doesn't collect it first.

When you reach the eightieth floor of the Q1 observation deck, the world's highest apartment block, you look across at the scenic rim, the Lamington National Park and Mt Warning. In the middle ground you see the Gold Coast airport and the mammoth structure housing the desal plant, a monument to the worst case of coastal overdevelopment in Australia. Perhaps the new slogan for Southeast Queensland could be: 'Head for the 200 kilometre city — carbon emission heaven'.

References

Australia Institute 2005, 'Greenhouse implications of proposed Sydney desalination plant', Webpaper 78, July.

Australian Journal of Politics and History (AJPH) 2004+, political chronicle on Queensland, available on the Wiley Blackwells journal website.

Brisbane City Council (BCC) Archives, news-cutting volumes on water, 1960–1995 (evidence of burial of water meters from an interviewee who cannot be named until retirement).

Brisbane Institute 2003, *Greenspace Audit of Southeast Queensland*. Four-page brochure by Gum, K. and Spearritt, P. and accompanying articles in the *Courier Mail*.

Cole, J. 1984, *Shaping a City: Greater Brisbane 1925–1985*, Brooks, Brisbane.

Cooley, H., Gleick, P. and Wolff, G. 2006, *Desalination, with a grain of salt: a California perspective*, Pacific Institute, June.

Dixon, Nicolee 2005, 'Grey Water on Suburban Gardens', Queensland Parliamentary Library paper, August.

Interviews 2007–08: A series of interviews was conducted with well-placed scientists within the Department who do not wish to be named until retirement, including some who have been formally retired but re-hired on a consultancy basis during the water crisis. Governments throughout Australia now regularly prevent public servants from speaking frankly about issues on which they are expert, especially if the issues are regarded as politically sensitive. Informed public debate suffers as a result.

Pretty, Lynda 2006, 'Water Restrictions in Queensland', Queensland Parliamentary Library paper, November.

Sinclair, Knight, Merz (trading as Queensland Infrastructure Pty Ltd) 2007, 'Environmental Impact Statement, Traveston Crossing Dam', October. This 1600-page report is available for download at www.qlwid.com.au and on CD rom. Quotes from chapter two: 22–4)

Troubled Waters: Confronting the Water Crisis in Australia's Cities

Spearritt, P. 2003, 'Can Brisbane remain a sub-tropical city', *Queensland Review*, vol.10, no.2, November: 25–36.

Spearritt, P. 2004, 'The 200 kilometre city: Noosa to the Tweed', 12-page brochure, Brisbane Institute.

Vuuren, Kitty van 2007, 'The Local Press and the Water Crisis', seminar paper, Centre for Critical and Cultural Studies, University of Queensland.

Warren, Matthew 2007, 'Politics of Water', *The Australian*, 13 October.

36

Chapter 3

Down the gurgler: Historical influences on Australian domestic water consumption

Graeme Davison

Australian city-dwellers are a thirsty people. Between the mid-nineteenth century and the present, the average daily consumption of water in Melbourne and Sydney has nearly trebled, from around 100 litres per head to around 300 litres per head. Industrial and other non-domestic users absorb about one-third of the flow but most is consumed in the bathrooms, laundries, kitchens, gardens and swimming pools of private homes. Usage has fluctuated across this period in a stepwise fashion. The first step in the later nineteenth century was associated with the introduction, under the influence of sanitary reformers, of piped water and underground sewerage, and had already pushed consumption to almost 200 litres per head by 1890, much higher than contemporary British cities. Even then it was clear that the Australian city-dweller's demand for copious supplies of clean water was inspired by a range of climatic, aesthetic and hedonistic, as well as hygienic, motives. 'More water is required because of the climate', observed William Davidson, engineer in charge of Melbourne's water supply in 1889; 'people bathe more here than at home, and another thing is, Melbourne is built very differently than any town at home, in that the whole of the settlement is on allotments with gardens' (Dingle and Rasmussen: 29). The second step occurred after the Second World War when suburban sprawl, rising affluence and advances in domestic technology pushed per-capita consumption to an all-time high, around 400 litres per day. And the third step, this time downwards, came after the 1980s as per-capita consumption was curbed by drought, increased prices, the adoption of water-saving technologies and more stringent regulations. Currently, under Stage 3 restrictions, Melburnians use about 280 litres per day.

Running water has long been regarded as an indispensable to any civilised community, but it is more indispensable for some purposes than others. From drinking and cooking, through showering, bathing and washing to flushing, watering and swimming, the household uses of water descend through an implied 'hierarchy of needs'. For some purposes, such as drinking, there is no substitute for water, while for others, such as disposing of human excreta, there are alternatives, such as the pan system (dunny can) or composting toilet. Our present ways of using water are a product, not of primal needs, but of history. They have been shaped both by culture (tastes, fashions, perceptions of health, virtue and comfort) and by path dependency (the particular array of technologies, governmental and pricing regimes we have created to supply and use water) (Shove 2003). By excavating the history of these arrangements, we are better able to think about how they might be changed or improved. The main reason that water usage in Australian cities is now unsustainable is not, however, that patterns of consumption have changed, but that urban populations have grown beyond the capacity of the catchments, which are themselves now subject to more variable patterns of rainfall. Environmental responsibility does not require us to return to some more virtuous pattern of past usage, for not everything about the past was virtuous and, in any case, the past is past and beyond recall. But it does require us to rethink the nature of our dependence on water, and to imagine how we might use it better. In this respect, history is an aid to imagination, if not a source of ready-made solutions. In this paper I consider, in turn, the changing patterns of water consumption for flushing, bathing, washing clothes, and outdoor uses, especially for irrigation and recreation.

The pursuit of health and morality

'Cleanliness', the eighteenth-century evangelist John Wesley famously declared, 'is next to godliness.' His adage reminds us of the strong link between Protestant morality and modern habits of cleanliness (Bushman and Bushman 1988). Clean water, applied inwardly and outwardly, was both an instrument and symbol of Victorian morality. In perhaps the most famous Victorian fable of cleanliness, Rev. Charles Kingsley's *The Water Babies* (1863), Tom, a poor chimneysweep, falls into a stream and drowns. He is magically transported into a kind of watery paradise where, freed from the cruel tutelage of his earthly master Mr Grimes, and washed clean from the soot that had once covered him from head to foot, he joins the happy company of water-babies. Kingsley's story draws, of course, on the religious symbolism of water as a medium of baptismal regeneration, but he was also an ardent supporter of sanitary reform, and concludes his tale by addressing his young readers with a more a mundane lesson: 'Learn your lessons, and thank God that you have plenty of cold water to wash in; and wash in it too, like a true Englishman.' Nothing, his young reader is assured, can go wrong 'as long as you stick to hard work and cold water' (Kingsley 1863: Chapter 8). A belief in the benefits of cleanliness and cold water became one of the pillars of a more general code of respectability, shared by working-class secularists as well as middle-class Christians. 'What is our doctrine?' pupils in the Lyceum, Melbourne's free-thought Sunday School, were asked. 'Frequent ablutions in cold water', was the reply (O'Dowd 1888: 16).

Conscious of this heavy overlay of Victorian morality, recent scholars have sometimes treated the fixation of contemporary reformers on dirt and cleanliness as an irrational fetish. Yet there was much in their recent experience to persuade contemporaries of the benefits of clean water. Between the early 1830s and the mid-1850s, Britain was thrice visited by epidemics of Asiatic cholera which together killed over 100 000 people. Cholera did not kill as many people as endemic diseases like typhus and tuberculosis, but its sudden onset, obscure causation and dramatic effects struck fear into the entire population. Victims were carried from their houses, writhing, sweating, vomiting and defecating uncontrollably, to over-crowded hospitals where doctors laboured, often vainly, to contain the epidemic. They were puzzled about the causes of the disease, some believing that it was carried directly from person to person, others that it was caused by miasmas or poisons in the air. Only in 1849, after the second major outbreak, did the London physician John Snow discover the correct explanation: that the cholera bacillus was transmitted through water supplies contaminated by human faeces. In 1853 Snow confirmed his theory by demonstrating that local victims of the epidemic had all taken water from the same contaminated pump in Broad Street, Soho (Longmate 1966: 201–11).

Flushing

Snow's discovery dramatically confirmed the importance English public health reformers already placed upon the need for a pure supply of water and the safe disposal of human wastes. The most influential of them, Edwin Chadwick, author of a famous 1842 inquiry into the *Sanitary Condition of the Labouring Population of Great Britain*, placed clean water and underground sewers at the top of his agenda. Like many of his contemporaries, Chadwick conceived the city as an organic system, analogous to the human body (Sennett 1994; Davison 1982: 364–6). A healthy city, like a healthy body, depended on the free circulation and exchange of vital fluids. Sanitary reformers sought to imitate the marvellous economy of the human body by integrating the supply of water, the disposal of sewage and the production of food in a single self-regulating system. Chadwick believed that piped water was as important for the safe disposal of human wastes through underground sewers as it was for clean drinking and washing. Combining water-supply and sewerage systems, he argued, would create 'an unseen, unostentatious, self-acting system of excretory ducts' (Chadwick 1965: 135 n.2). Chadwick's organic conception of the city, and its corollary, the interdependence of water-supply and waste disposal, deeply influenced the first generation of Australian public health officials many of whom, like Sydney's George Dansey, Melbourne's Tharp Girdlestone and Hobart's Robert Officer, had been trained in mid-century London (Mayne 1982: 58; Dunstan 1984: 244–6; Petrow 1995: 7–9).

The completion in 1865 of Sir John Bazalgette's massive underground sewerage system for London confirmed water-carriage as the preferred method of disposing of human wastes in towns. Designed for a city 10 times as populous and five times as dense as colonial Melbourne, it nevertheless remained the standard to which most Australian sanitary experts aspired. 'It is generally conceded that the sewerage or water-carriage system is the only one which collects and carries away the night soil and foul waters of a large town effectively', affirmed Melbourne engineer James Styles in 1888. 'Its action is prompt, and what is of equal importance, it is automatic. A substance dropped into any closet in Melbourne would not only be swept into a sewer at once, but it would be carried outside the city boundary in less than a hour.' (Styles 1888: 11) Underground sewerage was not just the best but, seemingly, the inevitable solution to human waste-management. Although it cost more than alternative methods of disposal, experts believed that its high cost would be justified in the long run (Girdlestone 1876: 12; Culcheth 1881: 191).

Between 1880 and 1910 both Sydney and Melbourne adopted underground sewers as the main means of disposing of human wastes. As the cholera had hastened the arrival of London's sewers, so the outbreak of typhoid epidemics in both cities during the 1880s had hastened the change. Sydney's sewers emptied into the ocean, Melbourne's to a large sewerage farm at Werribee beyond the city's western rim. Contemporaries welcomed the coming of the water closet to Australia's seaboard cities with relief, as at last removing a shameful blot on Australian civilisation. This was in spite of the fact that very few British cities other than London were any further advanced. 'At the end of Queen Victoria's reign, W.C.s were still unknown to the majority of her subjects', the British historian Anthony Wohl wryly observes (Wohl: 95).

Contemporaries probably exaggerated the benefits of water-carriage as a method of waste disposal. Historians who have closely examined the course of its introduction have sometimes wondered whether it was either efficient or effective. The economic historian W. A. Sinclair suggested that it was only when the costs of the old pan system exceeded those of underground sewerage that Melbourne opted for change. But his argument was convincingly challenged by David Merrett, who estimated the cost of the new system as actually twice that of the old (Sinclair 1975; Merrett 1977). The connection between improved health and the advent of the water closet was no clearer. While urban mortality, especially from typhoid and other contagious diseases, fell during the latter nineteenth and early twentieth centuries — the decades when Sydney and Melbourne were being sewered — the decline actually began before the inauguration of the sewerage system and was attributable, at least in part, to factors incidental to it, such as improved habits of personal cleanliness and street drainage (Dunstan 2003: 67–78). 'We are left with the fact that a major investment in public sewerage was established in Sydney and expanded, on what were essentially false premises', Dan Coward observes in his penetrating study of Sydney's environmental history (Coward 1988: 67). The decision to adopt underground sewerage was a momentous one determined not by economics or medical science — though both had some influence — but largely by the power of Chadwick's vision of a sanitary city linked by an 'unseen, unostentatious, self-acting system of excretory ducts'. Henceforward Australians, like Britons and Americans, would come to regard the water closet and the flush cistern as indispensable markers of civilisation.

Those who designed and built the sewerage systems were convinced that they would use no more water than the earth-closets and privies they replaced (Culcheth 1881: 190). However, between 1900, when the first houses were connected to the Melbourne system, and 1911 when most were connected, per-capita water consumption rose by almost 30 per cent, from approximately 50 to 65 gallons. Some observers put this down to the increased volume of water required for underground sewerage, but MMBW Chief Engineer William Thwaites was not persuaded. Not as much water was wasted as before in washing down yards and drains, he contended. Garden watering and extra showers in hot weather accounted for the increased consumption (Dingle and Rasmussen: 115). Yet there seems no reason why hosing and showering should have suddenly increased, and it may be significant that per-capita consumption levelled out after the system was completed.

In Melbourne water closets were flushed by pulling a chain connected to a three-gallon (13.6 litre) overhead cast-iron cistern; in Sydney the two-gallon (9 litre) cement cistern was standard (John Danks Catalogues, 1906, 1952). A contemporary who attempted an estimate of household water use in New South Wales allowed six gallons (27 litres) per head for water closets (Bruce and Kendall 1901: 55). In an era when the journey to the backyard dunny was longer, and every bedroom had a chamber pot, families may have held on longer, collected faeces and urine in a single vessel and flushed less frequently than today. The advent of the indoor loo from the 1920s and the multiplication of bathrooms and toilets in the post-war era finally banished the chamber pot, but the technology of flushing changed little until the 1970s when the Caroma company introduced the dual-flush toilet (Department of Environment and Water Resources). A standard dual-flush toilet uses less than half as much water as an old-style overhead cistern (3–6 litres per flush compared to 11 litres) and more recent 'smart-flush' models use even less (3 or 4.5 litres per flush) (Water Efficiency Labelling, 2007; Caroma website). Since nature presumably called as frequently in 1900 as in 2000 these technological improvements might have been expected to reduce the per-capita consumption of water for flushing, yet it is unclear whether they did so. Water closets currently account for between 15 and 23 per cent of household water consumption (around 50 litres) (ABS 2004 in Troy 2007), a similar proportion of per-capita use, but about twice the actual volume per head, as in the early 1900s (26 litres), if the rough, and perhaps inaccurate, contemporary estimates are accepted.

Bathing

'Cleanliness is the outward sign of inward purity', a guide to Australian etiquette advised upper-middle-class readers in 1885. 'Cleanliness of the person is health, and health is beauty. The bath is consequently a very important means of preserving the health and enhancing the beauty. It is not to be supposed that we bathe simply to become clean, but because we wish to remain clean' (*Australian Etiquette* 1980: 376–7). *The Australian Housewives' Guide* (1885), published in the same year for respectable working-class readers, offered similar advice, though in a more down-to-earth fashion. 'Personal cleanliness and neatness are the first requisites towards good housekeeping, and no woman who wishes to preserve her husband's affections, or make him comfortable, should ever waver in those attentions to her own person which will preserve whatever share of beauty she has' (*Australian Housewives' Guide* 1885: 72). In Australia, it was argued, daily bathing in cold water was 'a luxury all the year round', and a duty no self-respecting woman need fear: 'Have a good breakfast and you will be admirably strong enough to resist the slight shock of a cold shower bath' (Ibid.: 77).

Advice manuals written by stern moralists and reforming doctors may be a treacherous guide to the behaviour of contemporary city-dwellers, most of whom lacked the facilities, let alone the moral resolve, for the daily cold shower. Bathing and showering, as contemporaries understood those words, seldom meant full immersion in a deep tub or under a running shower. Most authorities recommended a daily sponge or hip bath, using only the few quarts of water that could be conveniently carried from the tap to the bedroom, together with a weekly warm soapy tub in order to open up the pores of the body and cleanse it from the impurities believed to accumulate there. The morning shower or bath was designed to refresh and stimulate; the weekly bath, usually taken in the evening, was designed to relax and cleanse (Muskett 1987: 24–34; *Australian Etiquette* 1980: 376–7; *Australian Housewives' Manual* 1885: 78; Wicken 1891: 194–7).

English visitors hailed the apparent superiority of Australian bathing arrangements. 'There is hardly the smallest cottage without its bathroom', Julian Thomas claimed in 1893 (Thomas in Dingle and Rasmussen: 29). He almost certainly exaggerated. In his fascinating account of *Our Home in Australia* (1860), the Adelaide artisan Joseph Elliott describes the contents of his house, minutely room by room. Only when he comes to the backyard, and mentions two washtubs stowed amidst hen coops and assorted rubbish, and characterises Saturday evening as 'ablution night', do we gain a glimpse of how the family performed what contemporaries called their 'toilet' (Elliott: 75–6, 78). Even in the 1880s many new houses designed for Australian workingmen had no bathroom. Those that had were tiny wooden enclosures tacked onto the back veranda, housing only a copper and a tin bath. A Queensland sanitary inspector considered the typical bathroom was 'the dirtiest room in the house' — sloppy, ill-lighted and smelling of urine, disgusting evidence that the bath doubled as a 'slop-sink'; that is, as a place to empty chamber pots (Elkington 1911).

The Ideal Bathroom of the 1920s, with its white tiles and chrome fittings, resembled a laboratory more than a boudoir. *Australian Home Beautiful*, 1925

26   THE AUSTRALIAN HOME BEAUTIFUL   November 12, 1925

A beautiful bathroom designed by Mr. C. H. Ballantyne, the Melbourne architect. It is tiled in blue and white, with a built-in radiator. An attractive feature is the slots in the tiled wall by the bath for soap, brushes, etc.

THE IDEAL BATHROOM
By F. L. KLINGENDER
A Melbourne architect describes the essentials of a small bathroom and some blunders for the homebuilder to avoid.

D Y the term "ideal bathroom" it is not intended to mean those palatial apartments which are frequently

east or north, thus getting the morning sun in winter and encouraging those Spartans who take a cold bath all the

Bathrooms were more common in middle-class than in working-class homes, though one should beware of assuming that cleanliness was a matter of class. In working-class Richmond, historian Janet McCalman found that bathrooms were more the exception than the rule. Many families lacked internal water supplies and were obliged to heat their bathwater in a wood-fuelled copper in the backyard. That so many did so, against such odds, showed how far ideals of personal cleanliness had permeated the respectable working class (McCalman 1984: 44). In middle-class Surrey Hills, where Moira Lambert grew up in the 1920s and '30s, the bathing arrangements were not much more advanced:

The daily ablutions were performed by heating water in a kettle, achieving the right temperature by adding dollops of cold, and then giving oneself a good wash all over from a larger enamel basin. Saturday night was bath night for the family, and I think that water must have been heated in the copper and toted in by bucket. Later we graduated to a gas bath-heater and shower, and finally — some time in the 1930s — a hot water service was installed (Lambert: 18–9).

New technologies influenced the evolution of the Australian bathroom, although they were themselves shaped by shifts in people's understandings of health, beauty, bodily comfort and pleasure. To understand them we need to appreciate how people *felt*, as well as what they *thought*, for the history of the twentieth-century bathroom is, above all, a history of the body and its senses, including that powerful stimulus to disgust, pleasure, arousal and nostalgia, the sense of smell.

In 1967 the artist Norman Lindsay, then approaching his ninetieth year, noticed a provocative article in the *Bulletin* magazine. The writer Sidney Baker had made the daring suggestion that the bushranger Ned Kelly may have been a homosexual, citing the fact that the famous outlaw was said to have used perfume and had danced with other men. Lindsay was indignant. 'Nearly all men of that era, irrespective of class, used perfume. My father, an Irishman, a horse-and-buggy doctor, and as dominant a male as ever wore whiskers, always finished off his morning toilet by dabbing his handkerchief freely with perfume.' Lindsay did not blame Baker for his ignorance of this fact, for he had 'forgotten it myself till I called it back to memory'. It was one of those 'trivialities which writers of the period rarely record, because they are conventions as understood by their readers'. In an era when open drains and reeking cesspools polluted the atmosphere, he explained, 'men, hurrying about their affairs, had no other resource but to clap a handkerchief loaded with perfume to their noses' (Lindsay 1990: 234–6).

People's sensitivity to smell, good and bad, not only changed over time, it also varied from one contemporary society to another. In his fascinating book *The Foul and the Fragrant*, the historian Alain Corbin (1986) notes the contrasting sensibilities of the nineteenth-century French and English.

The relative indifference shown by the French to cleanliness, their rejection of water, their long tolerance of strong bodily odors, and their continued privatisation of excrement and rubbish cannot be explained solely by a secret distrust of innovation, by relative poverty, or by slow urbanization. It was the collective attitude toward the body, the organic functions, and the sensory messages that governed behavior patterns. It is regrettable that historians have given scant attention to this somatic culture. (Corbin 1986: 173)

Towards the end of the nineteenth century, Corbin observes, the 'somatic culture' was characterised by the progressive 'deodorization of private space'. First in England, and later in France, the bathroom was being purged of unpleasant smells, and transformed into a 'sensually neutral and innocent space' (Corbin 1986: 175).

The crowded family bathroom was the bane of many post-war households, but a brake on water consumption. *Australian Women's Weekly*, 1951.

By the early twentieth century, Australians, too, had begun to transform the bathroom from a humble, often smelly, outhouse into a more 'innocent space' (Shove 2003: 94 and compare Lupton and Miller: 17–34). 'A well-equipped bathroom as an essential to the Home Beautiful whether it be Mansion, Villa or Cottage', advertisers assured readers of Australia's leading home magazine in the mid-1920s (*AHB* 12 February 1926: 5, 9). From its old place, amidst the copper and the laundry tubs on the back veranda, the bathroom was promoted into the most private and intimate zone of family life. 'The bathroom should be, as far as possible, close to the bedrooms, so that the occupants of the house do not have to undertake a lengthy pilgrimage in somewhat flimsy attire on a winter's morning', a Melbourne architect advised in 1925 (*AHB* 12 November 1925: 26; and compare Archer 1998: 116). In reinforcing a new sense of domestic privacy, the bathroom was, paradoxically, also becoming a mark of social status. 'Is your bathroom attractive? Can you show it with as much pride as you can your other rooms?' another advertiser inquired (*AHB* 12 February 1926: 9). Two themes dominate bathroom designs of the period: hygiene and comfort. Black and white tiles, enamel baths, pedestal basins and corrugated iron shower screens created an aura of antiseptic cleanliness, more akin to a laboratory than a boudoir. Yet the bathroom was also becoming 'the room in which comfort and convenience count most', especially to women whose image, often depicted in slinky satin dressing gowns, routinely accompanied advertisements for soap and bathroom fittings (*AHB* 1 December 1926: 10).

The pursuit of beauty and comfort, as much as the pursuit of health and cleanliness, gradually came to dominate contemporary attitudes to bathing. By the 1920s, the main dirt diseases, such as typhoid and dysentery, had been all but conquered (Cumpston 1927: 206–8). At the turn of the century advertisements for the most popular brands of bath soap, such as Lever Brothers' Lifebuoy, promised health and safety; but 20 years later, newer brands, like Lux and Pears toilet soaps, were promoted primarily as aids to personal beauty (*AHB* 2 September 1929; 2 December 1929: 83). Preserving one's complexion required a program of more frequent soapy baths. 'Frequent warm baths, at least one or two a week, are a necessity, with the thorough washing of the skin with a good soap to remove oily secretions, perspiration, dirt, dried skin etc followed by a brisk rub of the towel to complete the cleansing and stimulating process', a 1931 guide to housewifery recommended (Blackmore 1931: 6). 'Beauty with Pears' promised one advertisement, extolling the benefits of soap so pure you could almost see through it. From the 1930s Lux began a long tradition of advertising its product through endorsements by Hollywood movie stars. 'Keep that Wedding Day Complexion', promised Palmolive. Meanwhile, Lifebuoy reinvented itself by exploiting the fear of social ostracism and romantic failure associated with a dreaded new disease, 'BO' — or body odour.

Beauty, rather than hygiene, became the main theme of the bath soap advertisements in the early twentieth century. *Australian Women's Weekly*, 1951.

The most serious obstacle to the frequent warm soapy baths recommended by these health and beauty experts was the lack in most homes of a copious supply of hot water. 'Although the bath has always been acknowledged as an indispensable aid to health and beauty, the average home seldom has facilities for providing in sufficient quantity its necessary complement — Hot Water', a Melbourne manufacturer noted in the late 1920s (Danks catalogue, nd). He was extolling the benefits of a new invention, the wood-chip or gas bath-heater, as a source of 'instant hot water'. In practice, the hot water supplied by a bath-heater was rather less than instant or copious. In order to take a bath, one had to fill the heater, a metal cylinder containing sufficient water to half-fill the bath, light the gas-burner or wood-fire underneath, and wait 15 or 20 minutes until the clouds of steam issuing from the heater signified that all was ready. Turning on the hot and cold taps, the bather waited for the bath to fill before at last climbing in. The routine could be interrupted by exploding gas, wood-chips that failed to ignite, water that was either scalding hot or freezing cold, and impatient bangs on the bathroom door as other family members waited their turn. By the 1930s, a good chip bath-heater like the 'Kangaroo' ('a few chips of light wood give sufficient water for a steaming hot bath') sold for around £5. A superior brand, like the Braemar, which heated enough water for an eight-inch-deep bath in 25 minutes, cost around £10. The top of the range, the Triton Electric Bath Heater, costing up to £15, came closest to the Holy Grail of domestic comfort, instant hot water:

The gas bath-heater represented a modest but significant advance towards the goal of instant hot water. *Australian Home Beautiful*, 1928.

Hot-water system. *Australian Home Beautiful*, **1938.**

Page Four   THE AUSTRALIAN HOME BEAUTIFUL   March 1, 1938

Here Is The Hot Water System
for ...
Your Home

No matter how small or how large the establishment, there is an Electric Hot Water System to suit your requirements.

What a boon for the mother to be able to go and turn on the bath heater tap and steaming hot water is there!—without a moment's preparation, without a scrap of fear or of danger or of dirt; no chips to chop, no matches to strike, no escaping gas, no chance of suffocation, no ashes to clean up! (*AHB* 1 December 1927: 66)

Full-scale domestic hot-water services, heating water centrally and piping it throughout the house, had been advertised since the 1920s, but by 1943, when Professor Wilfred Prest and a team of social researchers surveyed housing conditions in inner Melbourne, only 2 per cent of households had a hot-water service and more than a quarter of households were still heating their bathwater on the stove (Darian-Smith 1990: 98). From the late 1940s, however, there was a rapid increase in the number of houses installing gas or electric hot-water services. Since the 1930s, the primary costs of hot water systems had declined in real terms and the introduction of off-peak rates for electricity further boosted demand. According to market researchers, Australian housewives considered 'continuous hot water' no longer a luxury, but a necessity (*AWW*, 23 March 1946: 13). By the early 1960s most Australian families could simply turn a tap to obtain a copious supply of hot water for washing, bathing, cooking and cleaning (Webber 2000: 175). Probably no other single innovation in the history of Australian domestic water consumption had such large effects. It heralded the rapid decline of the traditional bath-night, for now people could bathe or shower when they liked, without the exasperating wait outside the bathroom door as each tub of water was heated and filled. It laid the foundation for the introduction of other household appliances such as the washing machine and, later, the dishwasher. By removing barriers of time and discomfort, it also ushered in a new era of more extravagant water-use.

Melbourne and Sydney Cisterns

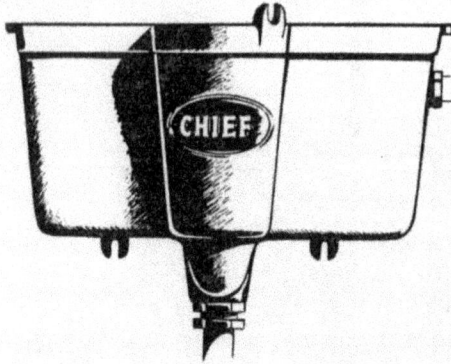

CHIEF
CISTERNS
U12.

Capacity 3 gallons, Melbourne House only.

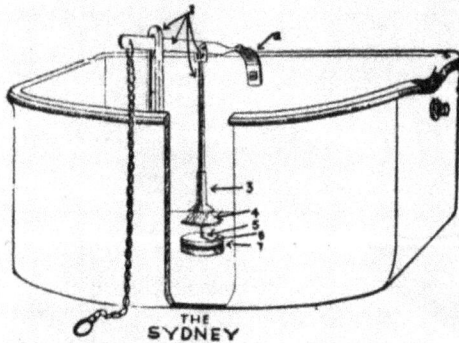

CEMENT
CISTERN
U13.

Capacity 2 gallons, standard Sydney pattern.

By the 1960s, plans for new project houses often included separate shower recesses and, from the 1970s, sometimes an ensuite bathroom as well (Garden 1995: 146, 150–1). 'There was a time when the bathroom was the worst decorated room in the house', a 1978 renovation manual began. 'No soft carpet here, no wallpaper, no daring colour scheme, no comfort; just a utilitarian, cold and sterile room where people washed' (Masters 1978: 7). Now, however, the bathroom was undergoing a 'revolution of thought. The bathroom has to have a touch of luxury, and why not?' Vanity suites, carpets, coloured wallpapers, and occasionally saunas and spas, were 'touches of luxury' in bathrooms that as yet usually remained structurally unchanged. Only in the 1990s did the implications of this 'revolution of thought' become fully apparent. The bathroom had always been a mirror of changing attitudes to the care of the body. From preoccupations with health and safety it had evolved, first, towards ideals of beauty and comfort, and, more recently, to the pampering of the body and the recuperation of the private self. A recent architectural writer conceives it as 'a haven for relaxation, grooming and total privacy' (Hasanovic 2005). Contemporary house designs now often incorporate three or even four bathrooms. The largest of them — typically the 'ensuite' adjoining the master bedroom —may be as large as a small bedroom in the houses of a previous era. Designed to express 'a sense of calm' and 'sensuous tactility', it may look through plate-glass windows into an enclosed courtyard (Hasanovic 2005). Showering, a ritual usually performed at the beginning of the day, is designed to stimulate the body to wakefulness after rest and may take only a few minutes. The bathroom, however, has now become a place for pampering the body and soothing the ego. The indispensable agent of this healing process — whether in sunken tubs of almost-Roman opulence or in pools, showers, spas and saunas — is an abundant supply of warm water.

Much of this advertising is (and was) directed at women and it is likely, as contemporary investigators have shown, that bathing and showering habits vary widely across the population according to sex and age (Randolph and Troy 2007; Hand *et al*. 2003; Gramm-Hansen 2005).

Since 1900 the amount of water used by the average Australian for bathing and showering appears to have roughly doubled. At the turn of the century a sanitary inspector's textbook noted that an average Sydney household consumed about 44 gallons of water per head per day. It reckoned 'that from 9 to 12 gallons of water per head [was] a necessity for a fairly cleanly existence, and this would not allow for baths to the extent desirable in this climate, especially in summer time' (Bruce and Kendall 1901: 55, 77). By the mid-1940s, per-capita consumption was approaching 70 gallons per head. Some of that increase was probably attributable to a modest increase in the frequency of bathing and showering, as advocated by health and beauty experts and made possible by the introduction of the bath-heaters. Only after the 1950s, however, when gas and electric hot-water services, and shower recesses, became standard fittings in most households, and the frequency of bathing and showering increased, did domestic water consumption rise to exceed 100 gallons a day. In the early 1950s, a family of two was reckoned to require about 20 gallons of water heated to 160 degrees Fahrenheit; a family of three, 28 gallons; a family of four, 35 gallons; and a family of five, 42 gallons, although some authorities recommended almost 50 per cent more (*Complete Household Guide*: 29; and compare *Ramsay's Architectural and Engineering Catalogue* 1949). By the end of the century, Australian households were using on average over a hundred gallons of water a day, over 20 per cent of which was for showering and bathing. This appears to be about twice as much as the members of present-day English households, many of whom, influenced in part by climate, maintain the traditional routine of weekly baths and daily sponge washes (Shove and Medd 2006: 5; and compare Hand, Southerton and Shove 2003). Australian visitors have long despised the English aversion to daily showering. Entering a London pub, Barry Humphries' caricature of the ugly Australian, Barry Mackenzie, exclaimed that he was 'as dry as a Pommy's towel'. But with the imminent onset of climate change, the Australian habit of the daily, or even twice-daily, shower may no longer be the virtue it once seemed to be.

Washing

For over 150 years, the rhythm of domestic life in Australia was defined by the weekly routine of household tasks. Monday — the first day after the Christian Sabbath, when the whole family was supposed to rest — was almost universally observed as washing day, and for women this was the most labour-intensive day of the week. In the mid-nineteenth century, the wash for a large family could occupy the washerwoman from early morning till well into the evening. The laundry for an average family could require, in washing, boiling and rinsing, as much as 50 gallons of water. Standard routines often involved the use of a copper and as many as three or four large tubs for washing, blueing and starching (Webber 1996: 151–2; Wicken: 177–87; Hackett 1916: 267–92). Working people were sometimes called 'the great unwashed', a title that, as the artisan Thomas Wright explained, signified that they had 'black hands to earn white money', presumably in contrast to the white-handed plutocrats whose money was not as clean (Wright 1868: viii). Their work clothes, soiled from their labours, often required thorough boiling, blueing, starching and rinsing to be made clean. Most people, however, owned many fewer changes of clothes than we do, and changed them less frequently. An expert on housewifery in the 1930s advised 'at least a weekly change of underclothes', a standard that most people would now consider low (Blackmore 1931: 8). Men wore business shirts several times, changing the detachable collars each day, before washing the shirt itself at the end of the week.

That this regime remained unchanged for so long was, in part, a measure of the low status that laundry, and those who performed it, enjoyed in Australian society (Webber 2000). At the end of the Second World War only 2 per cent of Melbourne households owned a washing machine and few people expected to do so (Webber 1996: 175). Then, quite suddenly, it all changed. By 1952, 28 per cent of Australian capital city households had acquired a washing machine, the most popular brand being the compact, inexpensive and freestanding Hoover (Opinion Research Centre 1952: 10, 12). By 1959 ownership had risen to 60 per cent; by 1963, 71 per cent; and by 1977, 91 per cent (McLeod 2007: 45). The standard washing machine then had at least twice the carrying capacity of the little Hoover and was almost completely automated. The advent of the washing machine also coincided with rapid changes in the mass production of clothing, especially the introduction of rayon, nylon and drip-dry or non-iron fabrics. Washing machine manufacturers now encouraged housewives to treat their washing machines as laundry baskets, throwing in clothes as soon as the wearers discarded them, and turning on the machine as soon, and as often, as they wished. In less than 20 years, the traditional Monday Washing Day had come to an end. The relative ease with which clothes could now be washed, dried and returned to the wardrobe removed a significant brake on the volume of clothes washed, and, in turn, on the amount of water consumed in the process. By the 1990s, the plunging price of mass-produced clothing, especially from China, and the gyrations of the adolescent fashion market further compounded the trend. Today, washing clothes accounts for approximately 15 per cent of domestic water consumption, a proportion that has hardly changed since 1900, even although the per-capita consumption of water has more than doubled (ABS 2004 in Troy and Randolph 2007; and compare Bruce and Kendall 1901: 55). The advent of the washing machine did not, of itself, greatly increase the amount of water used for washing clothes. Standard washing machine brands of the late 1950s used about 10 gallons of water for an 8lb wash, well under the 50 gallons used for the admittedly larger household of the 1850s (*John McIlwraith Buying Guide* 1959). The main effect of the arrival of the washing machine was not to wash the same clothes more efficiently, but to facilitate an increase in the size of people's wardrobes to accommodate the rapid changes of attire characteristic of a fashion-driven, consumer society.

Watering and wallowing

These shifts in the ways that families used water inside their homes paled, however, beside the extravagant, and increasingly volatile, consumption of water outside the home. During the 1970s, Melbourne water authorities calculated that while the average consumption of water inside the house had increased over the decade by 15 per cent, usage outside the house — for such purposes as watering gardens, washing cars and filling swimming pools — had increased by 52 per cent (Dingle and Rasmussen: 368). This was an era of rapid suburbanisation, when the quarter-acre block, the triple-fronted brick-veneer house and the Holden station wagon came to define Australians' conceptions of the good life (Davison, Dingle and O'Hanlon 1995: 2–17). The new suburbs of the 1950s and '60s were different from the terrace suburbs of the 1880s, and even from the bungalow suburbs of the 1920s. They were not only more extensive, with larger lots and bigger gardens, but they embodied new patterns of domestic life (Neutze 1977: 28–9). Houses were usually set back further on the block, with larger ornamental front-gardens; but the most dramatic changes, with the greatest influence on domestic water consumption, occurred in the space that, until that era, had usually been known as the backyard.

The arrival of the inexpensive Hoover washing machine was the first step towards the alleviation of the routine drudgery of the Monday washday and an economical consumer of water. *Australian Women's Weekly*, 1951

fits the family Purse !
does <u>all</u> the family Washing ...

The

HOOVER

ELECTRIC WASHING MACHINE

£43·10·0 *(or easy payments from 8/- a week)*

1. From sheets to socks it does the lot! All the week's washing for a big family !
2. Cuts out soaking, rubbing, boiling.
3. Washes whites in 4 minutes, silks and woollies in 1 minute.
4. Washes cleaner than you can by hand.
5. So gentle because the exclusive Hoover Pulsator doesn't yank and stretch the clothes but pulsates the water *through the weave.*
6. Cleverly sprung wringer takes even heavy articles with ease . . . makes broken buttons a thing of the past.
7. Tucks away in less than 3 feet of space when not in use . . gives you more elbow room in the laundry.

And, it's by the makers of the World's Best Cleaner

George Seddon has vividly described the functions of the suburban backyard, as he remembered it, in the 1930s.

> It had all or most of: a woodheap, often with a rickety woodshed with a low roof of galvanised iron and a fence for the back wall; a washhouse with two tubs and a copper, with a grate beneath it to heat the water and a wire rack to hold the Velvet soap and Reckitt's Blue; a clothes line; one or more tanks on wooden tankstands, with mint or parsley under or near the dripping tap in a cut-down kerosene tin, a dunny against the back fence, so that the pan could be collected from the dunny lane through a trap-door; there might be a kennel for the dog, though he often slept under the verandah; there was often a crude incinerator; often an old oil drum, although the rubbish was also burnt in an open bonfire. There might be chooks, usually in a chook house along the back fence, and sometimes a sleep-out, usually a verandah enclosed with fly-wire, but often free-standing (Seddon 1997: 153).

The backyard, Seddon implies, was essentially a utilitarian space: it was a site for those essential domestic activities that could not be accommodated within the house itself — growing food, storing fuel, washing clothes, harvesting rainwater, disposing of waste, human and otherwise, housing animals (compare Gaynor 2006). Already, in some Australian capitals, some of these functions, such as water-supply and disposing of human waste, had been outsourced or brought under the roof of the house, freeing the backyard for other domestic functions, such as housing and maintaining motor cars. Others soon became redundant: the woodshed fell victim to the gas space-heater, the washhouse to the washing machine, and the vegetable garden to the refrigerator and the supermarket.

The backyard, or 'back garden' as it was increasingly known, meanwhile began to evolve into a new kind of private domestic space shaped by aesthetic and recreational, rather than simply utilitarian, values. 'The backyard', John Murphy notes in his study of domestic life in the 1950s, 'was a place of pleasure, but also of retreat from the public world' (Murphy 2000: 27). Contemporary garden manuals characterise it as a zone of domestic seclusion, where the family could relax free from the prying eyes of neighbours and the stresses of everyday life. 'The back', a popular household guide advised, 'should be treated with the same care and thought as the front area. A pleasant garden in which to relax, or to be able to have one's meals in privacy, means much to the whole family' (*Complete Household Guide*: 60).

Modernist architects had begun to re-conceive the relationship between the suburban house and its surrounds. For more than a century, the architect and historian Robin Boyd observed, Australian houses had been designed 'to fight the un-English qualities of the Australian environment. The sun was shunned' (Boyd 1952: 93). From the late 1940s, however, modernist theory combined with a new appreciation of the environment to break down the old division between indoors and outdoors. In *Homes in the Sun* (1945), the architect Walter Bunning showed how suburban houses could be designed with L-shaped plans, large picture windows, deep eaves, stone-flagged terraces, pergolas and patios that brought the garden into the house. The ideal back garden should include both a 'Secluded Garden' for private relaxation and an 'Entertainment and Sitting Out' area, with a patio, barbecue or pool for social activity (*Your New Home Garden: it's* (sic) *design, cultivation and planning*, Lothian Publishing Company [1958]).

These changes implied a significant increase in the use of water for both irrigation and recreation. The most striking feature of the modern garden, consuming as much as 90 per cent of the water used for irrigation, was the lawn (Walsh 2004: 15). 'This is the age of garden lawns and the smallest home gardener accepts the lawn with all its maintenance as an essential part of the layout', *Your New Home Garden* proclaimed in 1958. Green grass had always been a desirable feature of Australian gardens, a nostalgic evocation of England and a symbol of the colonist's success in taming the Australian climate. 'Nothing can be prettier than an expanse of rich, green, close-mown lawn', *Coles' Australasian Gardener* noted in 1903. But without the services of a gardener to water and mow it, and a plentiful supply of water, the area of lawn that most householders could maintain was limited. The watering-can, aided by sparing use of the garden hose and sprinkler, and the push mower, were the suburban gardener's standard tools (*Yates' Garden Guide* 1918: 6–7; 1937: 7; 1941: 15). Typical suburban garden plans of the 1920s and '30s devoted most space to garden beds, paths and utility areas; lawn was often confined to the front garden where a circle of buffalo grass, the hardiest variety, linked paths and garden beds (Pescott: 9). From the 1950s, however, garden manuals began to recommend the use of rye, fescue and other finer, more water-absorbent varieties of grass and to make extensive lawns, with islands of garden and patio, a feature of their designs. Two technological developments — the nylon hose and the cheap motor mower — eased the task of lawn maintenance. Rubber garden hoses first appeared on the Australian market in the early 1900s (Blainey 1993) but they were prone to breakage and not everyone could afford one. 'A hose is a costly commodity, and the life of rubber is shortened considerably with ill-use', a 1916 gardening manual warned (*Searl's Ke* 1916: 24). The advent in 1945 of the Nylex plastic garden hose together with improved sprinklers and 'soaker hoses' offered suburban gardeners cheaper and more efficient methods of irrigation (Hewat 1983: 65–6). Even more momentous was the invention in 1952 by the Sydney engineer Merv Richardson of the Victa motor mower. This simple contraption, a two-stroke engine mounted on wheels and powering a rotary cutting blade, cost less than a quarter of the price of the older roller mowers. By 1961, Richardson had sold half-a-million machines and held 70 per cent of the market for motor mowers (Mason 2005). In less than an hour, the suburban householder could easily mow lawns that would have taken half a day, and considerable exertion, with an old push mower. The characteristic sounds of the post-war suburb were the drone of the motor mower by day and the steady beat of hoses and garden sprinklers, mingled with the shriek of children enjoying their cooling waters, on long summer evenings.

From the 1940s private swimming pools regularly appeared in articles on elite houses in the *Australian Home Beautiful*. They symbolised the life of luxury associated with Hollywood in the era of Esther Williams and Johnny Weissmuller. But few Australian families could afford the heavy costs of excavation, reinforced-concrete construction, tiling and filtering, chlorinating and pumping equipment required for a custom-built pool. The swimming heroes of the 1956 Olympics were products of the municipal baths rather than the private pool and during the 1950s and '60s thousands of local councils, energised by learn-to-swim campaigns, ensured that hardly a town or suburb in the nation was without its own 50-metre pool. The public pool represented a relatively economical use of water: on hot summer days hundreds of children and adults gathered to share what would later, in private hands, offer recreation to no more than a few families. The first changes came in the 1960s, with the advent of the mechanical excavator, ready-mixed concrete and the pre-formed drop-in fibreglass pool. Even more significant was the appearance in the 1970s of the easily erected 'above-ground' pools marketed by the Clark Rubber Company. 'A private swimming pool was once regarded as an expensive luxury seen only in the homes of the very wealthy', Ian Wigney, author of a popular guide to pool maintenance, noted in 1977. Now, he explained, a private pool was 'well within the reach of the average wage-earner' (Wigney 1977: 7–8). Already there were an estimated 300 000 private pools in New South Wales alone. There are no comprehensive statistics of swimming pools or of associated water usage in this period, but it is fair assumption that the rapid proliferation of private swimming pools, spas and water features, along with more extensive irrigation of private gardens, contributed strongly to the 50 per cent increase in outdoor use observed by water authorities during the 1970s. Today in Perth, the only capital with official statistics, about one household in five has a swimming pool, a proportion unlikely to be exceeded in any other capital, except Brisbane (ABS 2004).

The Victa mower eased the labour of garden maintenance and, indirectly, boosted the consumption of water in Australian gardens. *Australian Women's Weekly*, **1956.**

Conclusion

Per-capita consumption of domestic water peaked in the 1980s. Under the combined effects of 'user-pays' pricing, regulation (for example, restricted-flow shower heads, dual-flush toilets, drip-irrigation systems, water restrictions, and so on), public-awareness campaigns, and a prolonged drought, usage has since declined, although not enough to dent the underlying problem of water insufficiency.

Australia's current domestic water-using regime is a product of long-term and short-term, technological and cultural influences. Some of these may be easily modified; others — and not just the technological ones — may be changed only with difficulty. One of the effects of urbanisation and of the adoption of large-scale engineering systems of water supply and sewerage was to make less invisible the connections between the behaviour of water users and the natural systems on which they ultimately depend. This disconnection may have been reinforced by the adoption of market systems of delivery that tended to commodify water, encouraging users to expect that supply should simply expand to meet demand. Only when we recognise the historical and cultural forces that have shaped our present patterns of dependence on water for drinking, washing, watering, flushing and swimming, and institute cultural practices, technologies and feedback mechanisms that inculcate habits of sustainable water-use are we likely to ameliorate the present crisis.

The adoption of water-carriage as the main method of sewage disposal, for example, was the product of ideas characteristic of the mid-nineteenth century, but the assumption it has created, that human-waste disposal should be 'unseen, unostentatious [and] self-acting', dies hard. Any move towards a more environmentally sustainable system of waste disposal, such as composting toilets, would not only have to modify a massive infrastructure of underground pipes and pumping stations, but overturn a deeply engrained 'somatic culture' of odour, purity and danger.

Other domestic practices, such as the habit of daily showering, were shaped by more recent technological changes, notably the availability of 'instant' hot water, and cultural shifts, especially of hedonistic preoccupations with bodily comfort, privacy and self-care. Could bodies be pampered by less environmentally wasteful means? Or does the solution lie in technological fixes, such as water recycling, or the installation of monitoring devices such as shower-clocks? Showering and bathing is, by its nature, the most private form of water consumption, and hence the least open to external monitoring and control, although shorter and less-frequent showers would both reduce consumption and improve health, especially through the prevention of skin disease.

The consumption of water for clothes-washing may be reduced somewhat by the adoption of front-loading water-efficient machines, although the volume of washing is probably determined much more by the size of people's wardrobes and the rapidity with which they ring changes of garments from day to day and even hour to hour. Any move to modify this pattern logically begins, not in the laundry, where water is used, but in the department store, where clothes are bought, and in the nation's bedrooms, where decisions are made about to what to wear and when.

Policies to develop more-sustainable water-use in Australian cities have concentrated on the most visible site of water consumption, the suburban garden. It is easier to monitor use outside the home than inside, and the large amounts of water used for gardening and swimming, especially in cities with low rainfall and high average temperatures, seem to offer more scope for conservation than activities like bathing, showering and washing, which stand higher on the city-dweller's 'hierarchy of needs'. Outdoor water consumption varies considerably across Australia's cities, from approximately 173kl per household in Perth to 73kl per household in Sydney (State of the Environment Report). High rainfall and high population densities both probably play a part in Sydney's lower outdoor water consumption. Whether policies favouring greater urban concentration elsewhere would produce more-sustainable patterns of urban water-use, however, is much more doubtful. There is little evidence that flat-dwellers actually use less water per capita than residents of traditional family dwellings (Troy and Randolph 2007). Furthermore, any economies in water consumption from urban consolidation would be likely to appear only slowly and would have to be set against the reduced opportunities for water recycling on larger lots, the increased stormwater run-off from the greater area covered by impervious roofs, drives, roads and yards, and the subtle changes to micro-climate brought about by the loss of vegetation. Cities, as Edwin Chadwick was among the first to recognise, are complex systems in which the causes and consequences of human actions are manifold and often contrary to our expectations. There are no shortcuts and panaceas for the water shortages that have now become endemic in Australia's cities.

References

Aitken, R. and Looker, M. 2002 (eds), *The Oxford Companion to Australian Gardens*, Oxford University Press, South Melbourne.

Archer, J. 1998, *Your Home: The Inside Story of the Australian House*, Lothian, Port Melbourne.

Askew, L. E. and McGuirck, P. M. 2004, 'Watering the Suburbs; distinction, conformity and the suburban garden', *Australian Geographer*, vol. 35, no. 1, March: 17–37.

Australian Bureau of Statistics 2004, 'Domestic Water Use in Western Australia, 2004', 4616.5.55.001.

Dent, J.M. 1980, *Australian Etiquette or Rules and Usages of the Best Society in the Australasian Colonies*, People's Publishing Company Melbourne 1885, reprinted 1980.

Australian Greenhouse Office 2004, 'Minimum Energy Performance Standards Swimming Pools and Spa Equipment', Report no 2004/12.

Australian Flower and Vegetable Growers' Handbook: A complete and concise guide on gardening under Australian conditions, Carroll's, Perth 1931.

Australian Home Beautiful, 1920–40.

Australian Women's Complete Household Guide Illustrated, Herald Colorgravure, Melbourne, nd [early 1950s]

Australian Women's Weekly, 1940–60.

Bailey, J. 1992, 'Cleansing the Great Unwashed: Melbourne City Baths' in Graeme Davison and Andrew May (eds), 'Melbourne Centre Stage: The Corporation of Melbourne 1842–1992', *Victorian Historical Journal* 240, October 1992: 141–53.

Bedwell, S. 1992, *Suburban Icons: A Celebration of the Everyday*, ABC Books.

Blackmore, M. A. [1931], *Southern Cross Housewifery for Use in School and Home*, Auckland and Melbourne.

Blainey, G. 1993, *Jumping Over the Wheel*, Allen and Unwin, Sydney.

Boyd, R. 1952, *Australia's Home: Its Origins, Builders and Occupiers*, Melbourne University Press, Melbourne.

Browne, G. 2005, 'Girdlestone, Tharp Mountain (c.1823–1899)', *Australian Dictionary of Biography*, Supplementary Volume, Melbourne University Press: 145–6.

Bruce, J. L. and Kendall, T. M. 1901, *The Australian Sanitary Inspector's Textbook*, Sydney.

Brunning, L. H. 1949, *The Australian Gardener*, 30th edition, Robertson and Mullen Ltd, Melbourne.

Bunning, W. 1945, *Homes in the Sun: The Past, Present and Future of Australian Housing*, Baker and Company, Sydney.

Bushman, R. L. and Bushman, C. L. 1988, 'The Early History of Cleanliness in America', *Journal of American History*, 74, 4, March: 1212–38.

Carter, W. H. 2006, *Flushed: How the Plumber Saved Civilization*, Atria Books, New York.

Chadwick, E. 1842, 'Report on the Sanitary Condition of the Labouring Population of Great Britain' edited by M. W. Flinn, Edinburgh University Press, 1965.

Chandler, D. and W. 1939, *General Hardware Catalogue*, nos. 43, 48, 51 (Miles Lewis Collection, Melbourne).

Cleanliness is Next to Godliness: Personal Hygiene in New South Wales 1788–1901, Historic Houses Trust, Sydney, 1985.

Corbin, A. 1986, *The Foul and the Fragrant: Odor and the French Social Imagination*, Berg, Leamington Spa.

Coward, *Dan Huon Out of Sight: Sydney's Environmental History 1851–1981*, Department of Economic History, ANU, 1988.

Culcheth, W. W. 1881, 'The Drainage of Melbourne', *Melbourne Review*, 1 December: 185–93.

Cumpston, J. H. L. 1927, *A History of Intestinal Infections (and Typhus Fever) in Australia 1788–1923*, Australian Government, Canberra.

Darian-Smith, Kate 1990, *On the Home-Front: Melbourne in Wartime, 1939–45*, Oxford University Press.

Davison, Graeme 1983, 'The City as a Natural System: Theories of Urban Society in Early Nineteenth century Britain' in Derek Fraser and Anthony Sutcliffe (eds), *The Pursuit of Urban History*, Edward Arnold London: 349–70.

Department of the Environment and Water Resources, Design for Environment–Caroma Dual Flush.

Dingle, Tony and Rasmussen, Carolyn 1991, *Vital Connections: Melbourne and its Board of Works 1891–1991*, McPhee Gribble, Melbourne.

Douglas, M. 1984, *Purity and Danger: An Analysis of Concepts of Pollution and Taboo*, Routledge, London.

Dunstan, D. 1984, *Governing the Metropolis: Melbourne 1850–1891*, Melbourne University Press, Carlton.

Dunstan, D. 2003, '"Rules of Simple Cleanliness": The Australasian Health Society', *Victorian Historical Journal*, vol. 74, no. 1, April: 67–78.

Elkington, J. S. C. (ed.) 1911, *The Queensland Sanitary Inspector's Guide*, Brisbane.

Elliott, J. 1984, *Our Home in Australia, A Description of Cottage Life in 1860*, Flannel Flower Press, Sydney.

Elliott, W. 1903, *Cole's Australasian Gardening and Domestic Floriculture*, E. W. Cole, Melbourne.

Evans, I. 1986, *The Federation House: A Restoration Guide*, Flannel Flower Press, Sydney.

Garden, D. 1995, '"Type 15", "Glengarry" and "Catalina": The Changing Space of the A. V. Jennings House in the 1960s' in Graeme Davison, Tony Dingle and Seamus O'Hanlon (eds), *The Cream Brick Frontier: Histories of Australian Suburbia*, Monash Publications in History no.19, Clayton 1995: 140–53.

Gaynor, A. 2006, *Harvest of the Suburbs: An Environmental History of Growing Food in Australian Cities*, University of Western Australia Press.

Girdlestone, T. M. 1876, 'Under the Floor: A Lecture', Australian Health Society.

Gram-Hanssen, K. 2005, 'Teenage Consumption of Cleanliness'. Paper presented at the conference 'Kitchens and bathrooms: Changing technologies, practices and social organisation — implications for sustainability', June, University of Manchester.

Gresswell, D. A. 1890, 'Report on the Sanitary Condition and Sanitary Administration of Melbourne and its Suburbs', Melbourne.

Hackett, Lady [Deborah] (ed.) 1916, *The Australian Household Guide*, Wigg and Sons, Perth.

Hand, M., Southerton, D. and Shove, E. 2003, 'Explaining Daily Showering: A Discussion of Policy and Practice', ESRC Sustainable Technologies Program Working Paper Series Number 2003/4.

Hasanovic, A. 2005, *50 Great Bathrooms by Architects*, Images Publishing Group, Mulgrave.

Hewat, T. 1983, *The Plastics Revolution: The Story of Nylex*, Macmillan, Melbourne.

Hopper, E. W. 1935, *My Secrets of Youth and Beauty*, Reilly Lee Company, Chicago.

Ideal Hot Water Supply, John Danks and Son Pty Ltd nd.[late 1920s?] (Miles Lewis Collection).

Ierley, M. 1999, *The Comforts of Home: The American House and the Evolution of Modern Convenience*, Clarkson Potter, New York.

John Danks and Son Catalogue, 1906, 1952 edition.

John McIlwraith Industries 1959, *McIlwraith Buying Guide*, October (Miles Lewis Collection).

Kingsley, C. 1863, *The Water Babies*, London: Chapter 8 (Project Gutenberg).

Lambert, M. 1990, *A Suburban Girl: Australia 1918–1948*, Macmillan, Australia, South Melbourne.

Lindsay, N. 1967, 'The Question of Ned Kelly's Perfume' (*Bulletin*, 18 March 1967) in Keith Wingrove (ed.), *Norman Lindsay*, University of Queensland Press, St Lucia, 1990.

Lindsay, P. 1981, *The Australian Gardener's Guide to Lawn Care*, Reed, Sydney.

Longmate, N. 1966, *King Cholera, The Biography of a Disease*, Hamish Hamilton, London.

Lord, E. E. 1858, *Your New Home Garden: its design, construction and planting*, Lothian, Melbourne.

Lupton, E. and Miller, J. A. 1992, *The Bathroom, the Kitchen and the Aesthetics of Waste; A Process of Elimination*, MIT List Visual Arts Center, Boston.

McCalman, J. 1984, *Struggletown: Public and Private Life in Richmond 1900–1965*, Melbourne University Press.

McLeod, A. 2007, *Abundance: Buying and Selling in Postwar Australia*, Australian Scholarly Publishing, Melbourne.

Masters, M. 1978, *Make the Best of Your Bathroom and Laundry*, Summit Books, Sydney.

Mayne, A. 1982, *Fever, Squalor and Vice: Sanitation and Social Policy in Victorian Sydney*, UQP, St Lucia.

Medd, W. and Chappells, H. (nd), 'Market Environmentalism, Inequalities and Domestic Water Consumption', paper submitted to *Local Environment*.

Merrett, D. 1977, 'Economic Growth and Wellbeing: A Comment', *Economic Record* 53, 149, June: 262–8.

Moorehead, A. (ed.) 1942, *The Australian Blue Book*, Blue Star Publishers, Sydney.

Murphy, J. 2000, *Imagining the Fifties: Private Sentiment and Political Culture in Menzies' Australia*, UNSW Press, Sydney.

Muskett, P. E. 1987, *The Art of Living in Australia*, Eyre and Spottiswoode, London 1893, reprinted Kangaroo Press, Sydney.

Neutze, M. 1977, *Urban Development in Australia: A Descriptive Analysis*, Allen and Unwin, Sydney.

O'Dowd, B. (ed.) 1888, *The Australasian Secular Association Lyceum Tutor*, Tyzack and Picken, Melbourne.

Opinion Research Centre 1952, 'Public Relations Survey Conducted for General Motors Holden', George Patterson Pty Ltd, February (Roy Morgan Archive).

Pescott, E. 1926, *Gardening in Australia*, Whitcombe and Tombs, Melbourne.

Petrow, S. 1995, *Sanatorium of the South?: Public Health and Politics in Hobart and Launceston 1875–1914*, Tasmanian Historical Research Association, Hobart.

Ramsay's Architectural and Engineering Catalogue, 1931, 1949, 1954–6 editions (Miles Lewis Collection).

Randolph, B. and Troy, P. 2007, 'Understanding Water Consumption in Sydney'.

Robertson, E. J. J. and Cobb, F. E. T. 1925, *The Health Inspector's Manual*, Melbourne.

Robinson, W. M. 1960, *Notes on Laundry Work*, Angus and Robertson, Sydney, 1960, 1962.

Searl 1916, *Searl's Key to Australian Gardening*, Searl and Sons, Sydney.

Seddon, G. 1997, *Landprints: Reflections on Place and Landscape*, Cambridge University Press, Melbourne.

Sennett, R. 1994, *Flesh and Stone: The Body and the City in Western Civilization*, Faber, London.

Shanks Sanitary Ware, Flinders Lane Melbourne Catalogue no. 26, 1 July 1926 (Miles Lewis Collection).

Shove, E. 2003, *Comfort, Cleanliness and Convenience: the social organization of normality*, Berg, Oxford.

Shove, E. and Medd, W. 2006, 'The Sociology of Water Use', UK Water Industry Research Paper Final Report.

Sinclair, W. A. 1975, 'Economic Growth and Wellbeing: Melbourne 1870–1914', *Economic Record* 51, March: 153–73.

Small Swimming Pools, Cement and Concrete Association of Australia, Sydney 1961.

Styles, J. 1888, 'Lecture on Sanitary Reform for Greater Melbourne', Williamstown.

Timms, P. 2006, *Australia's Quarter Acre: The Story of the Ordinary Suburban Garden*, MUP.

Troy, P., Randolph, B. 2007, 'A New Approach to Sydney's Domestic Water Supply Problem', paper to 2007 State of Australian Cities Conference, Adelaide.

Troy, Patrick, 'The water services conundrum', unpublished paper 2007.

Mason, J. 2005, 'Turning Grass into Lawn The Story of VICTA as recalled by John Mason', John Mason, [Sydney].

Walsh, K. 2004, *Waterwise Gardening*, 3rd edition, Reed, Sydney.

Water Efficiency and Labelling Standards (WELS) Scheme, http://www.water-rating.gov.au/products/index.html

Webber, K. 1996, Romancing the Machine: The Enchantment of Domestic Technology in the Australian Home, 1850–1914, PhD Thesis, University of Sydney.

Webber, K. 2000, 'Embracing the New: A Tale of Two Rooms' in Patrick Troy (ed.), *A History of European Housing in Australia*, Cambridge University Press, Melbourne: 86–106.

Wicken, Mrs. 1891, *The Australian Home: A Handbook of Domestic Economy*, Edwards Dunlop and Co., Sydney.

Wright, T. 1868, *The Great Unwashed by a Journeyman Engineer*, Tinsley Brothers, London.

Yates' Garden Guide: for Australia and New Zealand, Arthur Yates and Co., Sydney, various editions.

Chapter 4

Nature, networks and desire: Changing cultures of water in Australia

Lesley Head

'We are in the middle of a desert and we get rain twice a year if we're lucky for an hour-and-a-half at a time.' (Keith, Alice Springs)

'We're not short of water here.' (Tom, Alice Springs)

'I love water [laughs]. I think I have a fetish about watering gardens ... I just get extreme pleasure out of being in the garden.' (Jacqui, Wollongong)

I have recently been analysing urban Australian relations to nature through the lens of the backyard garden. The quotes above are from three study participants, interviewed by team members during fieldwork in 2002–3. Water surprised us by emerging as one of the most important aspects of people's domestic environmental engagements. Nearly five years later, this consciousness is less surprising and likely even more entrenched.

Keith is referring to Alice Springs' average annual rainfall of 286mm. He is apparently well attuned to the realities of living in the centre of the driest permanently inhabited continent. He has learned to live with nature, since it is nature that delivers his water. Tom, on the other hand, is talking about the unrestricted water supply provided by the Roe Creek borefield and piped to domestic houses. He understands well where his domestic water comes from, via the complex infrastructure tapping into the Mereenie aquifer outside town. This can be understood as an eco-socio-technological network (a hybrid, or assemblage, in Latourian terms). In such a network, nature is just one component, and then one that is only constituted by virtue of its relationships to other phenomena.

Is it nature-thinking or network-thinking that will serve us best in providing sustainable urban water supplies in the next century? In this chapter I argue that many urban Australians have a well-developed understanding of our low and variable continental rainfall patterns and the necessity of adapting to these conditions. This consciousness ('nature thinking') influences patterns of practice towards reduced water consumption in ways that policymakers would be pleased about. Examples of such practices are outlined here. However, because it represents a disconnect with the complex infrastructure of domestic supply in urban contexts, it is unlikely to be as resilient to changed conditions as 'network thinking'. In what follows, I demonstrate detailed understanding of, and intervention in, networks of water supply and usage in the context of the backyard garden.

In contrast to a number of other environmental issues which stimulate more polarised responses, a commitment to reducing water consumption was shared across the diverse study population and manifest in a variety of changed practices. That such indications were present several years ago, at the beginning of the current drought, suggests substantial underlying support for stronger government action on water.

However, for many people their aspiration towards water-saving practices is in tension with the pleasure derived from water, and their expressed desires for more watery environments, as exemplified in Jacqui's quote. Summarised around the concept of desire, this trend is explored here as a contradictory pressure to that of water conservation. It is exacerbated by the consumptive forces of capitalism.

Cultural geographies of water

In arguing that there is a significant cultural shift occurring, I am not discussing here the actual levels of water consumption. Rather this is a complementary perspective that seeks to understand everyday practices and habits, and the processes that reinforce or change them. I am influenced by Shove's argument for a shift in the focus of social environmental research 'so as to comprehend the collective restructuring of expectation and habit' (Shove 2003: 4). Through a detailed focus on everyday practice, Shove shows, for example, how changes in what is considered 'normal' with regards to personal cleanliness and laundering have implications for water and energy consumption. Everyday knowledge and practice is an important issue for water managers in urban areas, with, for example, garden use accounting for 25 per cent of all household water use in the Greater Sydney area (Sydney Water 2003), and over 60 per cent in Alice Springs.

In bringing to awareness 'routinised' habits and interactions, retrieving them from the wordless background of 'practical consciousness', and subjecting them to scrutiny and reflection' (Sofoulis 2005: 448), such research provides an important complement to more quantitative analyses of both attitudes to and consumption of water (for example, Kolokytha et al. 2002; Nielsen and Smith 2005; Hurlimann and McKay 2007; Zhang and Brown 2005). As Sofoulis (2005: 448) argues, 'who normally entertains an attitude about a tap, a drain, or a sewage pipe?' Yet it is precisely everyday objects such as these that connect consumers and householders to the wider socionatural networks that constitute 'waterscapes'

(Swyngedouw 1999), so understanding habits of interaction with taps, pipes and buckets provides a crucial analytical link. Further, an emphasis on everyday practice can throw light on contradictory behaviours such as observed differences between attitude and practice (Askew and McGuirk 2004; Sofoulis 2005: 446), unrealistic perceptions by consumers of their actual water-consumption levels (Kolokytha *et al.* 2002: 399) and the use of discursive strategies to justify or excuse environmentally damaging practices (Kurz *et al.* 2005).

The study contributes to a growing body of work examining commonalities and differences in cultures of water (Strang 2004; Allon and Sofoulis 2006; Jackson 2006; Gibbs 2006). The theoretical framing draws on moves within geography and elsewhere to go beyond ideas of nature and society as separately constituted entities. New conceptualisations framed around hybridity and networks, as articulated for example by Latour (1993), Swyngedouw (1999) and Whatmore (2002), provide lines of approach to the complex entanglements of humans and nature, and to earth-surface processes pervaded by human agency. In an age of accelerating urbanisation, some of the most stimulating work illustrates ways in which cities are themselves saturated with non-human nature, and enmeshed with non-urban landscapes through intricate networks for the transfer of goods and services (Cronon 1991; Swyngedouw 1999; Gandy 2002; Braun 2005; Heynen *et al.* 2006).

I build particularly on the work of Kaïka (2005), who has provided an important study extending analysis of the modernist urban denial of nature to the space of the home, using the example of water. She argues that:

> [T]he social construction of the Western (bourgeois) home as an autonomous, independent, private space is predicated upon a process of visual and discursive exclusion of undesired social (anomie, homelessness, social conflict, etc.) and natural (cold, dirt, pollution, etc.) elements ... while the familiarity of the bourgeois home is dependent upon the visual exclusion of social and natural processes, the very creation of the safety and familiarity of the modern private home is nevertheless predicated upon the domestication of natural elements (water, air, gas, etc.) through a socio-economic production process. (7–8)

Kaïka makes the point that while the processes of social exclusion in and around the home have been extensively studied, for example in Sibley's (1988, 1995, 2001) influential work on socio-spatial classifications and boundaries, the exclusion of nature and socio-natural processes have not been adequately researched or documented (52). The above studies draw in turn on the work of anthropologist Mary Douglas (1966). In illustrating how different cultural groups order the world, Douglas argued that the classification systems (albeit themselves all different) leave certain things not belonging. In different ways, these come to be labelled 'dirt'; that is, disorder, or matter out of place. Kaïka argues that:

'Natural elements are not in fact kept altogether outside the modern home; but rather are selectively allowed to enter after having undergone significant material and social transformations, through being produced, purified, and commodified' (Kaïka 2005: 64).

Thus water is purified to become 'good' nature before it enters the house, and once it becomes 'bad' nature, in the form of sewage, it must not only be removed, but be visually excluded. In fact, of course, both the purified water and the sewage are hybrid forms dependent on complex material and social networks. The familiarity and comfort of the bathtub or swimming pool, Kaïka argues, require those networks to remain invisible, and the space of the home to remain clean and pure.

A number of recent studies have analysed 'droughts' as complex events in which rainfall scarcity, public discourse, changing regulatory regimes, technological networks and private behaviours are entangled (Nevarez 1996; Haughton 1998; Bakker 2000; Kaïka 2003). Full discussion of these wider networks in Australia is beyond the scope of this paper, but our fine-grained focus here on household behaviours provides important points of intersection with these other studies. Consumer resistance to water-conservation measures, and continued expectation of water as a 'naturally' abundant good, has been documented in cases where there is a lack of confidence in a privatised supplier (Haughton 1998: 426; Bakker 2000: 16) or a discursive disconnect between the householder and the networks of technology and supply (Strang 2004; Kaïka 2005). In Strang's analysis, the combination of privatisation of supply, water technology that encourages visions of an unlimited resource and increasingly individualised social lives has created a situation where, 'domestic users are … impervious to efforts to conserve water' (208). As a site where these networks are rendered partially visible and with which people engage on a daily basis, Australian domestic gardens provide a contrasting example; they are both arena and agent of changing practice.

Context and methods

The broader study is of 265 backyards and 330 backyarders (a number of couples were included) in Alice Springs, Sydney and Wollongong (Head and Muir 2006, 2007). Our sampling strategy was designed to encompass the socioeconomic and geographic variability in each of these areas (Commonwealth of Australia 2002). Participants were recruited through media advertisements and appeals, letterboxing, snowballing from other participants, and by liaising with community groups. Each backyard was visited and a semi-structured interview undertaken on site with the participant/s by one of a team of three researchers. Questions related to the activities of different members of the household, changes that had occurred over time, people's feelings about the space, what sorts of plants and animals were considered to belong, wider environmental attitudes and practices, and major influences. None of the initial questions was explicitly about water, but water emerged consistently in conversations about a variety of topics. The backyard was mapped and photographed, and checklists on the demography of the household, the structures in the backyard and the biogeography were completed. The interviews were transcribed and imported into the qualitative data-analysis program, N6. Initially, all water comments were content-coded for the context in which they were talked about and the practices described. Using a discourse-analysis framework, we coded for different kinds of motivations and investments.

Water and other 'environmental' issues

The period of fieldwork, 2002–3, corresponded to a time of significant drought onset in south-eastern Australia. By the end of winter 2002 there was considerable discussion of the drought in the Australian east-coast media. However, people were already talking about water well before this, and it seemed to us as interviewers that the drought sharpened an existing consciousness rather than creating a new one. Media influences in relation to water consumption and the drought were diverse and pervasive during this period. Messages about water consumption came through all forms of everyday media, even down to the reporting of dam levels on the TV news. With the imposition of water restrictions there was extensive advertising in the daily press, as well as mail-outs to individual households.

In contrast to the diversity of their views on other issues (for example, the role of trees, the importance of native species, love and hate of lawns), recognition of the importance of water conservation was the nearest thing to a shared environmental commitment across the study population. While a few thought 'the government' or 'they' should have built more dams, none contested the idea that as a society we need to change our ways when it comes to water. This is consistent with Kurz *et al.*'s (2005) finding in Perth that 'water was constructed as being a finite, precious and shared resource that must not be wasted' (616), in contrast, for example, to energy resources.

Water as nature: The arid-continent consciousness

Water is threaded through Alice Springs participants' responses with greater frequency than in the coastal interviews in Sydney and Wollongong, relating not only to the supply of water but also to detailed observations of weather and rain events. Comments about how much or how little rain has fallen were made in over 80 per cent of Alice Springs interviews, compared to 35 per cent for Sydney and 45 per cent for Wollongong. For example, Keith's teenage son Matt commented: 'This year we've had three millimetres since January 1st, nearly six months. In six months we've had three millimetres.'

A number of the coastal participants also referred to the aridity of the continent as an influence on their water-consumption patterns. These references tend to be spatially removed from the person's current location, and lack the detailed rainfall references of the Alice Springs participants. They are perceptions of a more generic, continental awareness of aridity.

> I think a major, major, major issue with the Australian environment is water. I think most of us don't really accept the fact that this is a really, really dry country and we use water as if it's really abundant and that's going to be a big issue for the next ten, hundred years, who knows. (Sue)

'The biggest problem this country has is the lack of water', said one woman, who went on to connect her present water-saving practices to a childhood on the land and the normality of scarcity. The connection to rural or agricultural childhoods and living with tank water was common.

For many of the Alice Springs participants, particularly those who have moved from elsewhere, it is the rarity or lack of rainfall events that acts to disrupt customary patterns, leading them to re-evaluate their consumption and consider alternative strategies:

> We first noticed when we moved here, people don't have gutters and when it rains there's all this water going everywhere, big sheets of it. It was amazing that people didn't have bigger tanks and collect the water. But water's cheap here and it's from an aquifer. It's non-renewable. It's about 20 000 years old. (Brad)

> I think I calculated that about three centimetres of rain will fill up the tank ... which is a typical summer downpour here ... And also if we can harvest some of the water that is falling in our backyard and use it, that's just saving what is actually a non-renewable resource out here essentially, which is the Mereenie bores which have only got 20 years left of water in them at the current consumption rates. (Michael)

Several coastal people related awareness of the harshness of the Australian environment to a more specific experience in their lives.

> Dave and I went travelling around western Victoria and NSW on a motorbike before we had kids and there were a lot of areas out there that were badly affected by drought ... I was totally shocked and just seeing animals that were dying in paddocks, and I can still recall the smell, it was just so bad. And I think we came back here and I think we were just like "that's amazing", we just take it for granted so much and we are living in the driest continent so we're looking at water tanks for the front and the back and for recycling as much water as we can. Yeah, and I think even when the drought breaks, I think we'll continue doing it. (Maureen)

We travelled across the Tanami [Desert] last year and I gained a sense of the real fragility; it gave me such a deep sense of kind of touching almost the womb of the land and realising how fragile it is, how precious things like water is and we're looking at a way to put water tanks in. (Maggie)

These stories demonstrate direct links between a specific life experience and a willingness to change consumption patterns. Both Maureen and Maggie seem to have used that experience to 'come to terms' with a dry Australia, but the connections are totally symbolic. The connection between the Tanami and Sydney in terms of water is, in a material sense, far-fetched. Using water tanks in Sydney will not save water for the Tanami. There is no strong relationship between water availability in the two places, either in terms of where the rain falls or where the storage and distribution infrastructure moves it to. A similar symbolic power is in operation when Barb tells her teenage daughter in the shower to 'save some water for the farmers, Jess'. She is expressing a broad consciousness of the arid continent rather than a belief that if Jess showers for less than 20 minutes in Sydney, the farmers in western New South Wales will actually get the water.

A further dimension of considering water as nature, or a purely natural resource, is in descriptions of changed behaviour in response to drought. There were many examples of this among the coastal participants:

Well I don't believe in watering the garden in the summer months when there's a particularly bad drought and quite a bit of the front lawn died this year completely. The buffalo died out completely there so I thought 'well, why maintain it?', because being light sandy soil there, the moisture drains out of it very quickly. (Ted)

But once the building work was finished I was going to put the lawn in but that was in the middle of the drought last year; it was last July so I just thought it was really silly to try and put a lawn in with the dry weather so it looked pretty horrible for months and months. (Trudy)

This is not to deny the power of the symbolic connections or of the practices altered to adapt to an observed scarcity of water. However, it is appropriate to question how deeply such practices are likely to be embedded. Will Jess lengthen her showers when the farmers are in flood? Will Ted and Trudy return to watering their lawns once it rains? The idea of water as pure nature is expressed in broader community debates that focus on the abundance of 'wasted' or 'unutilised' water in Australia's north, with naïve and simplistic suggestions to transport it southward (or transport agriculture northward).

Networks of water: Abundance or scarcity?

In Kaïka's (2005) reading of urban environments, the networks of water supply are hidden from or ignored by domestic consumers until something goes wrong. Yet both on the coast and in Alice Springs there is often quite detailed knowledge of these networks, whether for the drainage of excess or conservation of scarce water. A number of people explicitly visualised the pipes that brought the water to different parts of the house and garden, and recognised the implications for conservation.

> Because our ensuite is right at the front of the house, you can use two-and-a-half litres of cold water before you get your hot water through. So we trap that water as well. The same at the sink here at the back. It's just the set up of the tap. You turn it on. You hear the water coming through. You do what you have to do, you turn it off and it keeps on running. So again we've got a bucket in that sink and we trap all that water. For quick rinsing and stuff like that I just rinse my hands in that. So you get four litres of water in no time. (Robert)

> I would take a big bucket up and put it in the shower and when you turn that tap on, I mean your hot water has got to come from the hot water service which in our case is sitting there in the corner of the garage. So you've got a couple of gallons that comes off before the water is hot and then I would carry it down and water the garden or water something in the garden. (Mrs Heywood)

An important reason that people have detailed knowledge of the networks is that they are active agents within them. Participants recounted both creative and banal strategies to conserve and reuse water: the jug beside the sink, the bucket in the shower, the basin of vegetable-washing water, letting the lawn go, not planting annuals, water-saving shower heads, rain-soaker crystals, mulch and water tanks. 'Water-gathering' is the term we use for a loosely defined set of practices that were informal, irregular or unstructured in nature and differed from participant to participant.

> Even from ... the washing machine I tend to collect the water with buckets, you know, for the garden, water the trees and the lawn ... At the moment because of the drought we haven't had much rain for many months so even when I have a shower I have a bucket underneath it to collect the water and water the garden. The other thing is even when I wash the vegetables in the house or I wash my hands I tend to have a bowl or a basin underneath it to collect all the water. (Emily)

For Emily and a number of people, responses to the drought built on longstanding practices based on an ethic of not wasting. Several elderly

Macedonian women in Wollongong shared a generational practice of collecting water in buckets for use on their extensive vegetable gardens.

In Alice Springs, over 80 per cent of participants discussed the supply of water from the borefields, and the fact that it has a limited lifespan, although they had a range of views on how long it would last. In these conversations participants reveal a depth of knowledge of the technological networks of provision, discussing pumping, depths, capacity, as well as capital costs, rates of consumption, and comparisons to other parts of Australia. Implicit in this detail is a concern for the amount of water used by themselves and Alice Springs residents generally, focusing on change and extending to discussions on current and future direction regarding the government's water strategies. As one participant said of the community as a whole, 'consciousness has definitely shifted'. An example of someone with a good understanding of the network is Tom, whose quote opened the paper. A long-term resident of Alice Springs, he is conscious of the challenges of living in an arid environment, while maintaining a level of comfort through the provision of shade and water in his backyard:

> But currently, Alice Springs, we're not short of water here. The aquifer that Alice Springs uses to supply domestic water isn't small. I guess it's just a cost factor. So the government and authorities here encouraged people to have water-efficient gardens. You can do that with drip irrigation and the right selection of plants and still have a nice green garden like we have without having a huge water bill. It's the cost of managing the demand for water in Alice Springs that's the issue. If people consume excessive amounts of water here, it's going to cost the government more to expand the borefield that they're using and put new bores down and harvest the water there. So they're trying to manage water use for conservation reasons and also for economic reasons. But there is no actual shortage of water. (Tom)

On the other hand, Keith's observations of rainfall outlined above do not necessarily translate into reduced water consumption when combined with a busy lifestyle: 'So who has time to garden? ... I like the no-maintenance garden, you know, press the button on the American-style sprinklers and that's about all you have to do, leave it on the set timer.'

Those participants who discuss the socioecological networks reflect on long-term changes, particularly towards increasing demand. An important part of any water network is 'government', who were frequently constructed discursively as poor managers of the precious resource of water. This enabled some to distance themselves from any responsibility to 'fix the problem'. For others, the relevant policy-makers were considered to be a long way behind community consciousness and preparedness to change. For example:

[T]he government at its policy level has decided that it doesn't want to put water restrictions on or actually charge people what the cost of the water is, because they are trying to encourage people to come, and it's part of the frontier mentality that we still have in the Territory. (Peter, Alice Springs)

If there is little incentive beyond a moral one for Alice Springs residents to conserve water, and more costs are incurred than elsewhere to establish water-storage infrastructure, the water consciousness is even more striking. It indicates that both nature-thinking and network-thinking are strong, in different combinations. Indifference was rare. On the other hand, the diversity of understandings about exactly how sustainable the Alice Springs water supply is reflects a particular requirement of good network thinking; that its multiple connections and pathways be well understood.

Dilemmas and desires

Backyarders articulated a set of sensual and embodied engagements with water. It is a part of nature that is usually a source of pleasure, as illustrated first by discussions of the pleasures of watering, of which Jacqui's quote is an example. A number of women described a time of relaxation at the end of a day's work. This enjoyment of watering goes so far as to influence the watering systems they install, several describing deliberate decisions to not install drip-irrigation systems in at least part of their garden so that they could continue to enjoy hand-watering. Themes of pleasure, tranquillity and meditation came through in these conversations:

> I water a lot in summer and when I'm miserable I talk to the plants; I go out and let the plants cheer me up. And they tell me when they're thirsty or over-watered. (Betty)

> At least a few times a week I get out there in the morning and I water the garden. For me that's before I start my day and that is a very pleasurable activity, and as I water the different pots that are on the wall I check on the wellbeing of the plants just to see how they are travelling … and they're like my babies. And so I start my day with that uplifting experience and that's a major activity for me … I jog around the street, come back here and while I'm cooling down I'll water the garden and just check on the health of everything. (Patrick)

As Patrick indicated, this is a time when detailed observation of processes occurs. People do not just water; they observe the activities of ants and monitor the growth of plants. This is something that is lost if watering is an automated process in which the human does not have to participate. However, these are certainly not universal feelings. Jessica, for example, said: 'I hate watering. Some people love standing there with the hose and I hate it.'

In the Alice Springs context a strongly connected theme was shade, the necessary labour of creating a cool, green oasis. Long-time residents have watched the town change from the early days of bare-dirt backyards to a city with green lawns and automated sprinklers. Kerry, who has spent most of her life in Alice Springs, is aware that the changes in technology have created the demand for something that was previously not possible, but a past ethic of frugality has been usurped by habits and practices that are now well established and provide pleasure: '... we do keep getting told that it is not a renewable resource and it will conk out. But yeah, it's a bit hard to think in those terms when it's there. And the water is actually very good ... it's just so nice to do that, you water by hand.'

An extension of these pleasures is that participants voiced desires for more water in their everyday environments; swimming pools, ponds, streams, and water features were called on to bring serenity and the touch of water. Such desires are both fed and gratified by the lifestyle industry. Water is very clearly connected to visions of a nature that is tranquil and peaceful. In speaking about water features, people referred to beauty, the sound of running water, soothing natural sounds and the notion of creating a restful place within the garden: '[H]aving been in a city, close to the water, every day I passed the water and there's something tranquil and relaxing about that. Again, that's nature' (Diana). Justin described his swimming pool as not being about swimming, but 'about having water, being around water'.

The pleasures associated with water influence consumption in opposing directions. Hand-watering can increase or reduce consumption depending on how it is undertaken. If acted on, the set of desires focused on water features and swimming pools would increase the consumption of water. Other areas of unease relate to desires associated with increased affluence. In this quote the desire for a swimming pool is a somewhat guilty add-on to a conversation about replacing lawn as a water conservation measure:

> But we don't want grass, we are very pleased that we don't have any grass too ... It just takes water in Alice Springs, says us who are about to build a swimming pool [laughs]. (Alice)

Conclusions

Clear understanding of the 'nature' part of our urban water supply is a necessary but insufficient condition of making it sustainable. All elements of the network also need to be made visible. Water is now an important issue in both consciousness and practice of suburban householders. In our broader study, water contrasts with a number of other environmental issues that stimulate more polarised responses. Commitments to reducing water consumption are manifest in a variety of changed practices, many of which are hidden in the rhythms of daily life and can only be unearthed using qualitative research methodologies. Such methodologies also allow contradictions to be brought to light. The strongest example here is that aspirations towards water conservation are in tension with the pleasure derived from water, and expressed desires for more watery environments.

The presence and the value of the garden is not coincidental in these practices. The backyard garden is not a passive backdrop against which pre-existing attitudes are played out. Rather it is in the relationship between house and garden that people see, understand and participate in the network of water storage and distribution. Their active engagement with these processes enhances their capacity to manage and reduce consumption. They know their own power and they understand where and how to make a difference. To the extent that the garden or favourite plants are particularly valued, they are willing to make sacrifices, and to inject their own labour into the water network. This may explain why recent per-capita water consumption in separate houses with gardens in Sydney is little different from that of apartments and units (Troy, Holloway and Randolph 2005). On the other hand, domestic gardens, like other parts of our living spaces, are also sites of desire and consumption where intentions can come undone.

There is little support in this evidence for the construal of gardens themselves as environmental problems, and considerable support for the idea that more localised strategies for water collection, storage and distribution are likely to garner more support and active connections than Big Water schemes such as new dams. The widespread evidence of willingness to change practices suggests that there is underlying support for stronger government action on water, provided it is done in a way that maintains and utilises these human connections. The different scale of analysis provided by domestic ethnography adds a broader range of potential solutions to the complex issues of sustainable urban water supply. The everyday, habitual nature of human engagements with the non-human world provides an underrated human resource of considerable potential in the necessary shifts towards more sustainable cities. It should be regarded with cautious optimism.

References

Allon, F. and Sofoulis, Z. 2006, 'Everyday Water: cultures in transition', *Australian Geographer* 37: 45–55.

Askew, L. E. and McGuirk, P. M. 2004, 'Watering the suburbs: distinction, conformity and the suburban garden', *Australian Geographer* 35:17–37.

Bakker, K. J. 2000, 'Privatizing Water, Producing Scarcity: The Yorkshire Drought of 1995', *Economic Geography* 76: 4–27.

Braun, B. 2005, 'Environmental issues: writing a more-than-human urban geography', *Progress in Human Geography* 29: 635–50.

Commonwealth of Australia 2002, *Sydney. A Social Atlas 2001*, Canberra: Australian Bureau of Statistics.

Cronon, W. 1991, *Nature's metropolis: Chicago and the great West*, Norton, New York.

Douglas, M. 1966, *Purity and Danger*, Routledge and Kegan Paul, London.

Gandy, M. 2002, *Concrete and Clay. Reworking Nature in New York City*, The MIT Press, Cambridge, Mass.

Gibbs, L. 2006, 'Valuing water: variability and the Lake Eyre Basin, Central Australia', *Australian Geographer* 37: 73–86.

Haughton, G. 1998, 'Private profits — public drought: the creation of a crisis in water management for West Yorkshire', *Trans Inst Br Geogr NS* 23: 419–33.

Head, L. and Muir, P. 2006, 'Suburban life and the boundaries of nature: resilience and rupture in Australian backyard gardens', *Trans. Inst. Br. Geogr. NS* 31: 505–24.

Head, L. and Muir, P. 2007, *Backyard. Nature and culture in suburban Australia*, University of Wollongong Press with Halstead Press.

Heynen, N., Kaïka, M. and Swyngedouw, E. (eds) 2006, *In the Nature of Cities. Urban Political Ecology and the Politics of Urban Metabolism*. Routledge: London and New York.

Hurlimann, A. and McKay, J. 2007, 'Urban Australians using recycled water for domestic non-potable use — an evaluation of the attributes price, saltiness, colour and odour using conjoint analysis', *Journal of Environmental Management* 8: 93–104.

Jackson, S. 2006, 'Compartmentalising culture: the articulation and consideration of indigenous values in water resource management', *Australian Geographer* 37: 19–32.

Kaïka, M. 2003, 'Constructing scarcity and sensationalising water politics: 170 days that shook Athens', *Antipode* 35: 919–54.

Kaïka, M. 2005, *City of Flows. Modernity, Nature and the City*, Routledge, London.

Kolokytha, E. G., Mylopoulos, Y. A. and Mentes, A. K. 2002, 'Evaluating demand management aspects of urban water policy — A field survey in the city of Thessaloniki, Greece', *Urban Water* 4: 391–400.

Kurz, T., Donaghue, N., Rapley, M. and Walker, I. 2005, 'The ways that people talk about natural resources: discursive strategies as barriers to environmentally sustainable practices', *The British Journal of Social Psychology* 44: 603–20.

Latour, B. 1993, *We Have Never Been Modern*, Harvester Wheatsheaf, Brighton.

Nevarez, L. 1996, 'Just Wait Until There's a Drought: Mediating Environmental Crises for Urban Growth', *Antipode* 28: 246–72.

Nielson, L. and Smith, C. L. 2005, 'Influences on residential yard care and water quality: Tualatin watershed, Oregon', *Journal of the American Water Resources Association* 41: 93–106.

Shove, E. 2003, *Comfort, Cleanliness and Convenience*, Berg, Oxford.

Sibley, D. 1988, 'Survey 13: Purification of space', *Environment and Planning D: Society and Space* 6: 409–21.

Sibley, D. 1995, *Geographies of Exclusion. Society and Difference in the West*, Routledge, London and New York.

Sibley, D. 2001, 'The Binary City', *Urban Studies* 38: 239–50.

Sofoulis, Z. 2005, 'Big Water, Everyday Water: A Sociotechnical Perspective', *Continuum: Journal of Media & Cultural Studies* 19: 445–63.

Strang, V. 2004, *The Meaning of Water*, Berg, Oxford.

Swyngedouw, E. 1999, 'Modernity and Hybridity: Nature, *Regeneracionismo*, and the Production of the Spanish Waterscape, 1890–1930', *Annals of the Association of American Geographers* 89: 443–65.

Sydney Water 2003 http://www.sydneywater.com.au/everydropcounts/garden/index.cfm (Accessed October 2003)

Troy, P., Holloway, D. and Randolph, B. 2005, 'Water Use and the Built Environment: Patterns of Water Consumption in Sydney', City Futures Research Centre, UNSW, Sydney: Research Paper No. 1.

Whatmore, S. 2002, *Hybrid Geographies. Natures, cultures, spaces*, Sage, London.

Zhang, H. H. and Brown, D. F. 2005, 'Understanding urban residential water use in Beijing and Tianjin, China', *Habitat International* 29: 469–91.

Chapter 5

Urban water: Policy, institutions and government

Steve Dovers

This chapter seeks to connect discussion of human behaviours around water not to taps, toilets and timing showers, or dams and desal plants, as much discussion (very usefully) does, but to the policy processes and instruments, institutional and governance systems, and household realities that shape human and organizational behaviours toward water in a modern society and economy. The focus is on urban water, but the discussion necessarily travels to rural water and issues like energy that cannot easily be separated from water. The chapter comprises a series of linked discussions on issues that surround more singular policy debates around water, hinged on the proposition that water policy is better constructed as being about far more than just water, and where the prospects for behavioural and institutional change become both more complicated and realistic.

The rising tide of debate

It seems too easy a question to ask why we are so worried about water in Australia today. The overall answer is scarcity — not just of available water, but of convenient supply options, opportunities for quick reform of infrastructure and institutions, resources both financial and informational, and capacity in the environment to receive wastewater. In the spring of 2007, parts of rural Australia are at breaking point in both house and paddock and perennial horticultural plantings may be abandoned along with small communities. Towns larger and smaller face shortages never before imagined. Cities face near-term restrictions, some inconvenience and cost, and are worried about an escalation of both.

On any international comparison Australians use water rather profligately: in rural irrigation systems, in industrial processes, at tourist resorts, on sporting fields and golf courses, and in houses and gardens. Against increasing scarcity, there is a reasonable expectation that there are ripe, low-hanging fruit in efficiency gains. Australians, at least urban ones, have never really been told to be frugal (the odd mild water restriction aside) but, rather, have been encouraged to splash it about in all sorts of ways. That is a hard legacy to shift and involves much more than changing immediate behaviours concerning appliances and orifices.

In recent years, water has become prominent in national political and policy debates in a manner unprecedented in Australian history, as a major issue at all levels of government, and through much stronger and more comprehensive national policy development in the form of the COAG-agreed National Water Initiative (NWI) and the more Commonwealth-imposed National Plan for Water Security. Why now? Drought, obviously; or, more accurately, a particularly widespread and persistent drought. That is still an incomplete answer. The slow and inexorable progress toward centralism in the Australian federal system is a major factor, sharpened by federal government of 1996–2007 but reflecting longer trends. This fulfils Deakin's prediction regarding the states being bound to the 'Chariot Wheels of the Commonwealth' made in 1902 with an insightful reference to drought (see Connell 2007). Centralism combines with populist political styles and the rise of Executive power to make big, sudden policy shifts and big, sudden infrastructure announcements more likely. While the broader, stable and moderating traditions of Australian governance and political confluence are apparent (Wanna and Weller 2003), in particular sectors such as water, instability and rapid change do occur. Concerns over climate change are influential, instilling an understanding of possible permanence of water scarcity. So too is the slow and incomplete move toward seeking ecologically sustainable development, mixing concerns over water supply with arguments for environmental flows and evidence of the downsides of the way in which we deal with wastewater. Increasing demands for participatory approaches to the management of natural resources influence the way in which water is understood and managed, although this is more obvious so far in rural than urban contexts. The marketisation of water services and agencies following the neo-liberal revolution has altered both water management and expectations of relative public and private benefits (for example, Sheil and Leak 2000; Gowland and Aiken 2003). Finally, there is the fact that the easiest option — increasing supply — has run up against the constraints of a flat, dry land. The easy dams have been built.

As with most eruptions of interest in major policy issues, there are multiple factors behind the current topicality of water, however dominated by a severe drought. This is not new. Economic scrutiny of the wisdom of unthinking bulk supply of water began in earnest with Davison's (1969) *Australia wet or dry?* Economic scrutiny, both sophisticated and simplistic, of investment in water infrastructure increased in intensity from the 1980s onwards, although large, panicked and arguably inefficient expenditures on engineering fixes have not ceased (although they have more in rural areas largely as a result of few remaining supply-augmentation options).

Water debates and policy activity, and, at times, real policy change, follow El Niño drought cycles with a slight time lag and depressing regularity. We do not here delve deeply into the broken past of water attentiveness — for example, post-war development-oriented reports and programs, and the 1963 and 1975 national water resources surveys — but take national water-policy debate and development and data gathering in recent times as an indicator. Water 2000, the most comprehensive set of reports and recommendations on water in the country's history, followed the early 1980s drought (DRE 1983) but soon faded from memory and influence. So did Water Review 85, the first time water resources and use were surveyed together nationally (DPIE 1987). There was to be a 'review 95' and each decade hence, but this basic need and pledge was washed away in a few wet years. The early 1990s drought led to 1996 election promises from both sides to examine water resources again, and the National Land and Water Resources Audit ensued (www.nlwra.gov.au), not matching too well with previous data sets or with the ABS's emerging Water Accounts, but very welcome nonetheless. Early 2000s drought has driven development of a new national data set, the National Water Commission's Australian Water Resources 2005 program, one that assumedly will be maintained consistently and improved, at least until the end of the NWI's implementation schedule of 2004–14 — an unusually long-term policy timeframe. Australia worries about and measures water when there isn't enough of it. Maybe this time the reality of the driest inhabited continent, and the most variable rainfall on earth, will sink in permanently rather than quickly evaporate.

Physical as well as policy activity has been lumpy in time. The bulk of Australia's water storages, rural and urban, were built in a rapid period from the 1960s (Smith 1998), in answer to multiple factors and needs — post-war development, rising populations, possible expansion of rural commodity exports, memory of previous droughts and particularly the 1940s in NSW, and especially the engineering and fossil-fuel-fired ability to build big things. That rush of dams is an infrastructure legacy that locked in and further fashioned the deep-seated water behaviours and institutional inertias that now present as problems. A hectoring focus on people's water-use behaviour ignores the fact that these behaviours are determined, enabled and constrained by the operating environment in which they take place — just as a focus on 'green consumerism' can deflect attention from deep-seated inconsistencies between the function of modern economies and ecological (and arguably social) sustainability. At best, a focus on individual consumption behaviour change ignores how a modern society functions; at worst it conceals a blame-shifting onto the individual that eases the need for effective reform of patterns of production and consumption, settlement and governance.

'Water policy': Distilling the intent

The focus of this chapter is not the fine-scale management of urban water or the detail of water consumption behaviours. Other contributions in this book do that. It discusses, rather, the settings that do much to determine management and behaviour — policy processes and interventions, institutional systems through which these are negotiated, and patterns of governance that surround these. We talk of 'water management' but it is really about managing people, whether individually or collectively in households, firms, communities and cities.

Some clarity helps. Water policy is in the news a lot, and policy interventions of all kinds are proposed. Policy interventions are always a form of social engineering to greater or lesser extent, and are thus *interventions* in society. This discussion follows the definitions of policy, institution and related terms used in Connor and Dovers (2004) and Dovers (2005). Policy interventions are intended to change human behaviour in order to further some social, political or policy goal, whether that goal is clearly apparent or widely shared or not. (Also, policy interventions almost always have unintended and multiple impacts, such as on water consumption, and some such will be noted as we proceed.) Whether a tax incentive is used, an educational campaign or strict regulation, a subsidy to a water-efficient technology or any other specific policy instrument — the point is to encourage, enable or enforce behaviour change on the part of individuals, households, demographic strata, firms, professions, communities, organisations or governments themselves. Policy instruments are *messages*, conveying information whether in the form of a threat, exhortation, appeal to generosity, mildly suggestive signpost or blaring claxon (Dovers 2005). The strength may vary but the intent does not: a harsh and confronting advertisement is a strong message, just as strong as a strict regulation and hefty fine or a large tax impost.

It should always be remembered that policy interventions in urban water seek to drive some very widely distributed and highly personal behaviours embedded in daily lives and close environments — washing bodies, cooking and cleaning up afterwards, going to the toilet, creating pleasant backyards to live in. Changing behaviours is serious business, and doubly sensitive and difficult when it gets personal. The most apparent urban impact of drought from a local government management point of view is grass, in parklands and especially on sporting fields and in swimming pools. Recreation and sport may seem to some trivial, but are hugely important socially, culturally and economically, and in answering their local democratic imperatives, do not doubt that local government councillors know this very well.

To decide not to make a policy intervention is a message also, confirming existing behaviours, as does a supply-side response (dams, desalinisation plants, groundwater tapping) that does not interfere with and indeed encourage continuation of use patterns. That is again a conscious choice regarding human behaviour, even if apparently unthinking. This clarifies what water policy and management is about, and weighs against the all too common and simplistic debates around the relative merits of different classes of policy instrument — regulation doesn't work, education is the most fundamental approach, leave it to the market, etc. *What is the best medium for the message*; or what is the best mix of instruments to convey the message effectively, clearly and fairly? Here I will take a non-discriminatory approach to policy instruments, accepting all as possibly valid, without favouring, as many do, one class of instruments (regulatory, market, educational). Better to consider first the nature of the problem and the desired ends, and then consider the means (policy processes and instruments, technologies and management strategies, institutional forms).

Policy interventions emerge from policy processes shaped within institutional systems and by governance. At this higher level of organisation and abstraction, it is equally all about human behaviour. Institutions are how we manage and structure transactions in a manageable and orderly way. These transactions are social, legal, economic, formal and informal. Seeking sustainability is more than anything an issue of institutional change (Connor and Dovers 2004). Governance is the way in which the state, private sector, civil society and public interact to lead to decisions about institutional reform and policy directions — how human behaviours are managed. We undertake collective endeavours and reconcile differences (or fail to) through institutions and processes of governance — water policy is no exception.

Watering city and country

Water in Australia is largely considered in the separate domains of urban and rural. This split exists in the narrative of settlement, political discussions, the jealousies of pub talk, supply and management systems, and policy and institutional regimes. Rather than urban–rural differences when referring to households and water, the real split is reticulated or independent supply. The split is evident too in research. Relative to population and economic activity, far more resources are expended in rural water and related research in natural resource management (NRM) than in the urban domain, despite the fact that major urban centres contain some 85 per cent of the population, and the bulk of social and economic activity. Yet on spatial extent, ecological impact and gross share of consumptive and non-consumptive water use, the rural domain deserves attention. In national policy debates, in the NWI and the National Plan for Water Security, the greatest focus is rural — or, rather, the Murray-Darling Basin. Some researchers advocate the importance of one over the other and therefore, like the policy regimes, can ignore interactions. It would be better for researchers, policy-makers and the public to believe two things at once. Both rural and urban water (and extant, missing or proposed links between them) are important and deserving of close attention.

Nevertheless, the NWI does instruct a linking of rural and urban water management (regarding the NWI, see Hussey and Dovers 2007); and, increasingly, attention is being paid to water management in hitherto overlooked peri-urban areas. Mostly, the linkage is interpreted in terms of transfer of water, whether through trading or simple capture, from rural sources to thirsty cities, and the provisions have yet to be pursued with any vigour and consistency. But this is a future area of research and policy activity with strong behavioural dimensions that are merely hinted at below.

While the focus here is urban, we do need to maintain an integrated focus, or at least recognise the other domain as a reference point. It may be that, behaviourally and in policy terms, it is in smaller rural settlements, localities, farm households and the non-metropolitan Indigenous domain where the sharpest insights into the human dimensions of abstemiousness are to be found. In southern NSW, there are small settlements that have been beyond Stage 5 water restrictions for more than five years; in rural homesteads in drought-struck areas children are bathed in suspect water hauled manually from diminishing farm dams while generational legacies of homestead gardens are irrevocably dead; small-town tourist ventures have suffered massive turnover losses and entire communities have lost the facilities to play sport; in Indigenous settlements, water supply and quality are of third-world standard. It is not a discussion of how long one should stay in the shower, as turning on the tap is an empty gesture. Such situations are far beyond the experience and imagination of the vast bulk of urban Australians and bear reflection.

Later a simple characterisation of phases in urban water management will be presented, and it is foreshadowed that the earlier phases have not been replaced but still struggle for space amid multiple values surrounding water and the inertia and path dependency of agencies and institutions. The households in drought-struck grazing districts, technological changes aside, are close in their use of water to predecessors. Joe Powell's (2000) symbols of two fundamental black and white Australian water dreamings still apply — the Rainbow Serpent of creation and the Water Cannon of intervention. The formal recognition of Indigenous water values in the NWI in 2004 — a great advance even if yet to be addressed seriously — shows that the oldest Australian institutions regarding water, Indigenous law and story, have not gone away (Jackson and Morrison 2007). These two symbols are non-urban; later we will consider some purely urban icons of water use.

Energy, water and climate

Water debates are littered with 'stuff of life' arguments, and water is indeed a fundamental requirement for life, and this property adds a human-rights and basic-needs dimension to policy considerations. Not just human life — few of our water decisions do not have implications for the integrity of ecological systems and biodiversity. It is also a systemic economic resource, irreplaceable as an input to production and consumption, and thus implicated in countless other policy sectors: the stuff of life, and the universal solvent, lubricant, coolant and producer of steam.

For example, it is sub-optimal to seek to understand water and to make policy about water without factoring in another, even more systemic resource — energy — which is equally topical at the moment (Proust *et al.* 2007). Climate change threatens water systems, and is caused largely by energy use. The shower links major uses of water and energy in the household. Different water-supply options (solar, nuclear desalinization, gas turbines, and so on) have very different energy demands, and vice versa. High-efficiency irrigation systems are abstemious but use pumps and energy-rich products; water-wasting flood irrigation is marvellously undemanding on energy. Most pollution is related to our uses of water and/or energy.

This warns against hydrological determinism, of narrow water-fixations in policy and management. In the non-urban domain, recent topicality of water issues has diluted hard-won and still-evolving integrated catchment and landscape-management regimes, where water is one, albeit crucial, component of a portfolio of issues to be managed in a coherent fashion: soils, vegetation, production, biodiversity, etc. In cities, too strong a singular focus on water may serve to embed a forgetting of the links to related issues and trends. The fact that, on a cradle-to-grave basis, the great bulk of water use attributable to household end-point consumption occurs before final consumption (known as virtual or embodied water, used in growing, manufacturing and transporting goods and services) indicates how deeply embedded water is in a modern economy. The same applies to embodied energy.

Water policy is a cross-sectoral, -portfolio and often -jurisdictional matter. These attributes suggest that extant policy and institutional settings, fashioned around levels of understanding before much attention was paid to water (and energy) as sustainability problems, are *prima facie* likely to be inadequate (Dovers 1997). Water is obviously a cross-disciplinary issue, and while research attention has flowed strongly toward water in recent years, it is largely a portfolio of separate components. Less-than-satisfactory integrative, intellectual attention to water and energy issues stems in great part from the inability of the research and education community to rigorously traverse the disciplinary boundaries within which intellectual activity is clustered (very productively, too, in all sorts of ways).

Water policy is a narrow construct, then, as water use is firmly linked to and determined by other policy and management sectors: planning, landscape and catchment management, fire policy, building regulations, energy availability and price, the evolution of appliances, garden and leisure fashions, and so on. Glib as it sounds, everything is indeed connected to everything else; to be effective, water-policy interventions must take account of the links, of the knowledge that underpins them and the policy frameworks constructed.

Watering policy and institutions

What follows is a sharply summarised and simply characterised view of overlapping phases of water management, policy and institutional forms in Australia, which roughly accord with approaches taken to other natural resources such as energy, and many other policy problems (for example, Bolton 1980; Frawley 1994; Connell 2007).

Indigenous Australian water law and management, although non-urban, have the longest pedigree and argument for 'fit' to the Australian environment, on the basis of impressive longevity. It can be surely expected to contain — although this is shamefully unexplored — a great variety of geographically and culturally defined variation. The difficulty of translating traditional Indigenous water management to contemporary urban settings is immense, except in a general value-shift sense, but the reality of 50 000-plus years of human–water interaction cannot be ignored.

Early Australian urban-water management was characterised by (i) rapid development of an understanding of the variable Australian hydrological environment, and (ii) a mostly localised and ad hoc approach to capture, provision and disposal of water. As populations grew and production demands and technologies drove increased water use, issues of adequacy and safety of supply became apparent.

The rise of public-health considerations in cities — potable supply and safe treatment of human and other wastes carried in water — defined the imperative for development of reliable, widely (if not universally) accessible bulk supply and disposal of water and related wastes. Though nascent technologies could have prompted a policy of widespread independent water management at household level and in some industrial settings, there was instead a move toward large dams, bulk piped supply, and (later) big-pipe waste disposal and mass treatment of wastes.

An 'engineered ascendency' (Powell 2000) and institutionalisation of bulk water capture, supply and waste disposal was the result, beginning in the late nineteenth century and implemented vigorously. In major urban settings, independent statutory authorities (water boards) became highly professional, powerful and purposeful organizations, with a singular and highly effective focus. In small towns, the same overall approach was taken, with local government rather than state agencies taking responsibility.

Rethinking these Leviathans began in the 1980s and gathered pace in the 1990s, amid rising concerns around cost-effectiveness, organisation flexibility, environmental impacts and, above all, economic efficiency. Agencies were put on a more commercial footing through corporatisation (in mandate, form and style of title) and outsourcing or privatisation of ancillary functions. Some revision of supply-oriented mandates occurred, and marginal efficiency measures were pursued. Whether these revised institutional forms equal a significant shift in emphasis remains to be seen.

These 'phases' reflect a sequence of shifts in policy styles and institutional forms; however, they are not strict or exclusive. Indigenous water management is extant, and in some areas pure state provision exists alongside corporatisation. Overall, they are not inconsistent with broader trends in public administration and governance, in their social goals and the organised means of addressing these.

There have been few big shifts in the underlying institutions — the rules of the game — in the history of Australian water management, much policy and management turbulence aside. This is to be expected: institutions, as opposed to the organisation that manifests them, by definition, do not change often and only occasionally quickly. Australia's first environmental regulation affecting water came quickly when Governor Phillip acted on the belief that faeces were out of place in the Tank Stream, and much has happened since. But the rules of the games, basic institutional forms, are slower to change. There is considerable path dependency in institutions — we generally operate within an institutional system that is far more consistent with past rather than present knowledge and imperatives. This inertia resides in modes of understanding, statute law, organisational and professional expectations and norms, and physical infrastructure. Modern Australian cities have developed assuming abundant low-priced water and energy, and the urban structure, housing forms and transport options that abundance make possible. Those assumptions have only recently been questioned. With only around 2 per cent of the building stock changing each year, and major public and private infrastructure projects having functional lives of many decades, rapid change is difficult, especially in the myopia of modern politics (see Marsh and Yencken 2004).

In rural Australia, the shift from Indigenous management to British riparian doctrine was the first big institutional change, and the second occurred in the years around Federation consistent with Deakin's injunction to overturn riparian doctrine and invest control of water with the Crown. Little changed fundamentally until recent years when concerns over economic efficiency, environmental quality and limits to supply saw management of water shifted toward private control and property rights (including environmental) began to be refashioned. Whether this current phase of institutional change will be as profound as some hope remains to be seen (see various contributions in Hussey and Dovers 2007).

In urban Australia, white occupation was the first big shift, but the early years of establishing white governance saw largely ad hoc urban water management. The second institutional shift was the rise of large state agencies entrusted with water and sanitation, beginning in the late nineteenth century and becoming formalised behemoths of unquestioned authority and expertise. Importantly, the logic was not so much water supply, but *public health*, borne of the post-Industrial Revolution wave of urbanisation, microbial diseases, scientific understanding of disease, and *distributional equity*, to offer all citizens a safe and reliable supply of potable water. The third shift, not unlike in rural Australia, is partial and ongoing: the corporatisation of water-supply agencies, application of pricing mechanisms, and partial incorporation of environmental constraints. Consistent with neo-liberalism, the most significant and effective policy expression came from the National Competition Policy-inspired COAG water reforms of the 1990s (for example, Sheil and Leak 2000; Smith 1998; Cater 1998). The social and institutional logic shifted, although the shift was not widely understood. While public health and basic service provision were still important, *economic efficiency* of operations became an underlying imperative, given direction by the corporatisation of water agencies within statutory frameworks that gave greater primacy to financial return. Citizens became consumers. As traditional statutory authorities, these agencies were thought by reformers to be inflexible and wasteful (and certainly that critique held in some ways); the answer was to make them more like the idealised private firm. As with other manifestation of corporatisation and privatisation, much depends on the quality and comprehensiveness of the statutory framework, and whether public-good functions such as long-term monitoring, public health, infrastructure planning, and so on are catered for (for example, Richardson and Bosselman 1999).

This indicates a widespread issue in natural-resource management, and many other policy sectors, where significant institutional change can occur under the guise of policy and management changes that are perceived to be, or sold as, a relatively straightforward matter of instrument choice or organisational form. Market-oriented policy approaches, especially those that refashion property rights, are not simply 'another option in the policy toolbox', but are transformative policy options that carry with them institutional change (Connor and Dovers 2004). Property-rights instruments — individual transferable quota in fisheries, tradable water rights in water — shift the policy logic from distributional equity plus some ecological consideration, to economic efficiency plus sustainability concerns. Those impacted by a new policy regime, and indeed sometimes those proposing and implementing it, may not fully understand that crucial shift. Trouble usually ensues. Moreover, such market instruments are not themselves an end or a singular means, but first require support from other, co-ordinated policy instruments (statutory, communicative, informational), and second are pro-sustainability only as an efficient allocation mechanism within a robust sustainability (scarcity) limit. Also, the standard conditions for any functioning market — good information, clear property rights, reasonable symmetry, and so on — are required. This basic economics is often assumed away or left un-discussed in the establishment of markets in natural resources (and other things). This does not discount the potential of market mechanisms, but seeks clear recognition of their implications and design of policy regimes in a manner sensitive to this understanding. As proposals for and implementation of water pricing and trading increase, this should be borne in mind.

Urban Australia as we know it grew up with and became accustomed to reliable and abundant supplies of clean, safe water. This is a truly tremendous achievement, and a curse. The achievement is of urban amenity, quality of life, convenience and, above all, public health. The curse is an inflexible, institutionalised water and waste system based on a large-scale, engineered, 'big pipe in, big pipe out' logic. This system does not encourage frugality, is hard to refashion given the inertia of infrastructure, and does not readily admit independence of supply or inventive and less environmentally disruptive use of wastewater. Some of the best-quality treated water in the world is used for all applications, from drinking to gardening and flushing toilets, and huge quantities of it are required to shift wastes in low gradient, gravity-driven sewers to bulk treatment plants from whence valuable water and nutrients are ejected into the ocean (or, worse in some cases, rivers).

The organisations and professions that developed and oversee this at once wonderful and regrettable system of water and sanitation are often as resilient as the infrastructure. This is an expectable example of the co-evolution of professional and organisational norms. That appears to be as true of today's marketised corporations as it was of the statutory authorities from which they arose. In their defence, many of the things that are increasingly being demanded of corporatised water utilities are simply not rational for them to do within the statutory mandate that governments, representing society, have given them. This includes such things as mandating efficiency, undertaking long-term environmental monitoring, collecting water data that informs other than straight commercial accounting needs and investing in different forms of supply and treatment infrastructure.

That applies to individuals as well, hectored to save household water use but subject to instructions and constraints in other dominant areas of policy and technology that weigh against such frugality (see below). Inert and commonplace, the garden hose should be up there with the Victa mower and Hills Hoist as icons of suburban Australia. It symbolises the end point of the bulk-supply reticulated system (the real end point, the hidden sewer, is not an attractive icon): the hose as the serpent of waste in our gardens. The advent of unbelievably cheap black poly pipe and associated efficient drippers has revolutionised water use and efficiency in backyard and on farm, but simply is nowhere as satisfying to use. Or is the real serpent the spa bath? In future, the water tank or dry-composting toilet may become the symbols of urban water behaviour; currently in some rural towns it is a front yard of dust.

The many initiatives to get individuals to change their water behaviours are good and logical, but policy-driven behaviour change as we have noted is not a straightforward matter. Institutional change in a democratic society cannot be too discordant with public values, lived experience and thus supportive normative change. At a practical level, there are issues of how capable and motivated people in houses in Australian cities are when it comes to responding to demands for frugality and effort in managing their personal water consumption.

Real human water behaviours

Thus far, the connection between policy, institutions and governance and human behaviour has been discussed at a broad level. Now consider the realities of real human behaviours in real human settings in real time. Contemporary ideas in water policy and management do imply significant changes in human behaviour. So it needs to be considered what such changes entail, against the many other factors that impinge on individual and household behaviour. Picture an 'average' or at least a believably typifiable Australian household. It contains two adults, both of whom now work full jobs to maintain a house and debt, the value of which has been increased significantly in line with the logical impacts of behavioural changes driven by policy interventions in the housing finance system such as negative gearing. Finances are tight. They collectively work longer hours than they did 10 years ago and than their parents ever contemplated. With retrograde funding of public education, one parent is committed heavily to helping at the local primary school and the other manages a junior sports team; commitments that each take six–eight hours out of a crowded week.

Enter a serious commitment to using water and energy more sustainably, with an eye not just to immediate use but whole-of-production-chain use (embedded energy and water). Water tanks are expensive to buy and install and require maintenance (and the embodied energy in that metal or plastic!), and the reinstallation of all light fittings, small subsidies aside, costs money. Doing installation oneself is cheaper and might be rewarding, but who can find the time, let alone plumb it into the toilet? Managing the internal temperature manually via shutters and scheduled closing and opening of windows takes 10-times longer than using the air-con. Negotiating rebates and green-energy subscriptions seems to take even more time than managing the kids' mobile phone plans. Finding reliable information on the embedded energy and water in all those products is hard work amidst a deluge of commercial, government and NGO claims. Water prices do not really drive frugality and would need to treble to really stand out in the household budget: energy prices are larger, but the gains are hard to make to the extent that they really show dollar benefit. Composting the wastes is yet another everyday responsibility, and using greywater on the garden keeps slipping because a quick half-hour hit with the sprinkler is much easier. So it goes on.

Such constraints apply beyond the household scale. Many prescriptions for sustainability require time-intensive collaborative endeavours — landcare and waterwatch groups, neighbourhood gardens, clean-up days —and all when traditional and important voluntary enterprises such as charities, junior sport and school fund-raising struggle to find human resources and time. And we expect people to do more?

This is a crucial and often-overlooked point — more ecologically sustainable behaviours generally replace external sources of energy and other resources such as water with metabolic (muscular) energy and labour time, and the intellectual effort to organise and manage (for the fundamentals, see Boyden 1987). Conservatively, can we propose that the time-cost of running a considerably more sustainable household may be six–eight hours per week? That is, roughly, another day's working time. Interestingly, there appears to be a convergence when one anecdotally surveys the time-cost of serious non-paid work activities such as chairing a Landcare group, being an active P&C executive member, coaching a junior cricket side, or getting real about household sustainable management — about an extra day's work per week. Sure, down-shifting or dropping out might square the circle, reducing expenditure to match reduced income, but that is not currently a widely available option. Reliable part-time work is not is easy to find.

This is obvious stuff, and has been since the onset of the industrial revolution — fossil fuel and other external uses of energy extend and supplant the metabolic energy used to do things manually (Boyden 1987). That is the time-saving magic of each instance of using energy and technology, even if the cumulative impact adds up to a tail-chasing feedback between expense, work and time. However, the necessary changes in human behaviour implied by many prescriptions for water and energy frugality seem not to mesh at all well with household realities and weekly time budgeting.

Of course, it is not only people's time-scarcity that weighs against behaviour change — the conservatism and inertia of water- and energy-supply agencies, the locked-in patterns created by past urban planning, the constraints or at least lack of encouragement in building codes, an economic and taxation system that does not particularly encourage frugality, and of course — and perhaps most powerfully —the active anti-sustainability that is carried in the aggressive marketing in a consumer society. The daggy 'stop the drop' adverts in low-ratings slots are up against the flash, prime-time spa and air-con ones. Economic growth and water and energy use are tightly coupled. Those six–eight hours a week of pro-sustainability time and efforts entail flying in the face of the inevitable logic of the inherited urban system, market-defined social expectations, the inertia and discouragement of powerful organisations, and the growth imperative of the economic system.

So, small adjustments in financial and other policy settings to encourage frugality aside, what prospect for behaviour change at the individual and household level? The playing field is not even. Recall that not to act in a policy sense is a conscious act of ensuring that current behaviours exist. This means that, unless pro-frugality policy interventions significantly outweigh these multiple anti-frugality policy messages, then on balance there is no reason to expect other than changes at the margin.

Conclusion: The singular and systemic in water policy

The foregoing has traversed some issues around human behaviour and water use in Australian human settlements. Currently Australia is in a water panic, but faced by significant path dependencies and inertia in water and sanitation systems, statutory and organisational dimensions of the institutional system, and human behaviours both individual and collective. There are also countervailing policy and commercial messages weighing against the prospects of significant change to water behaviours. How long the current policy panic will last is unknowable, but there is a greater chance now, with a focus on climate change and variability, of attention being maintained for longer than in previous dry spells. Actually, things are fine in metropolitan Australia (although not in some smaller towns and rural areas), but we have no idea how the politics of really severe or even absolute (as opposed to the current inconvenient) scarcity would play out, when whole categories of use become either impossible or grossly unviable from an equity point of view. Things may not get so bad, and whether we should seriously prepare for such a state of affairs is a matter of individual risk-aversion as well as careful projection.

The current water panic is characterised by a contest between a range of singular policy and technological responses with very different behavioural implications. Large-scale supply-side options lessen the need for behaviour change, whereas significantly greater household responsibility for capture and storage of water and treatment of wastes implies large behavioural adjustments. In between is a mix of these, mediated by various technologies. Many specific policy interventions are proposed — regulatory, economic, communicative — but all have the same intent of changing human behaviours, which is not a trivial undertaking.

What is not discussed as much are meaningful shifts to other policy sectors so that all that determines water use is as consistent as possible. Or, should another social and policy goal be considered to override water issues, then the inconsistency in policy and technological messages are transparent. Water policy needs to be connected to other policy domains, the agendas and mandates of over agencies and portfolios. Attempts to alter individual and household behaviours would take account of the other determinants of those behaviours: technological, commercial advertising, countervailing financial measures, available urban form, energy issues, and more. That would require connections at present poorly developed in both the research and policy-and-management domains.

It would also require reviewing and revising the statutory frameworks and thus socially-mandated parameters of function of key organisations, and most particularly water utilities — both directly state-run and corporatised. Similarly, a wide and deep review of the statutory framework surrounding water management is overdue (for example, Fisher 2007). Serious attention to whole-of-chain water demand in systems of production and consumption would be needed, to address embedded as well as direct water use. Planning regimes, and especially strategic planning, might be reviewed for their consistency with water issues. In a pro-sustainability institutional system, strategic environmental assessment or some equivalent would be applied to major policy proposals to assess cross-sectoral impacts on environment, society and economy in a precautionary manner (Dovers 2006).

In all the above, it would be best to include broader sustainability imperatives rather than only water — the issues are linked, and not many reviews are likely to be possible. None of these reviews and institutional reforms will save a single drop of water; rather, they would reorder the operating environment of the policy intervener and their subject, whose behaviour is meant to change in the interests of greater clarity and possibility. They would treat water and related issues as complex and systemic, rather than simple and thus suited to singular responses. They would allow for strategic rather than silver-bullet responses to water scarcity.

In the 16 years since the integrated agenda of sustainable development (prefixed as 'ecologically' in the Australian term 'ESD') was first formally adopted in policy (UN 1992; Commonwealth of Australia 1992) and the two decades since the idea was first clearly articulated (WCED 1987), we have seen principles of ESD enunciated in countless policies and laws in Australia and internationally. However, serious policy attempts to implement it in sectors have been rare and partial — in Australia, Regional Forest Agreements and now the NWI are examples (for example, Dovers 2002; Dovers and Wild River 2003). In urban water, National Competition Policy has been more influential that ESD ideas, and carry a particular set of assumptions and reliance on market-oriented instruments. While we might well get by with disintegrated and ad hoc approaches to urban water, with the odd drought-induced catch-up, climate permitting, it is worth considering the merits of something more integrated.

References

Bolton, G. 1981, *Spoils and spoilers: Australians make their environment 1788–1980*, North Sydney: Allen & Unwin.

Boyden, S. 1987, *Western civilization in biological perspective: patterns in biohistory*, Oxford: Clarendon Press.

Cater, M. (ed.) 1998, *Public interest in the National Competition Policy*, Sydney: Public Sector Research Centre University of NSW.

Commonwealth of Australia 1992, 'National strategy for ecologically sustainable development', Canberra: AGPS.

Connell, D. 2007, *Water politics in the Murray Darling Basin*, Sydney: Federation Press.

Connor, R. and Dovers, S. 2004, *Institutional change for sustainable development*, Cheltenham: Edward Elgar.

Davidson, B. 1969, *Australia wet or dry? The physical and economic limits to the expansion of irrigation*, Melbourne: Melbourne University Press.

DPIE (Department of Primary Industries and Energy) 1987, *1985 review of Australia's water resources and water use*, 2 vols. Canberra: AGPS.

Dovers, S. 1997, 'Sustainability: demands on policy', *Journal of Public Policy* 16: 303–18.

Dovers, S. 2002, 'Sustainability: reviewing Australia's progress, 1992–2002', *International Journal of Environmental Studies* 59: 559–71.

Dovers, S. 2005, *Environment and sustainability policy: creation, implementation, evaluation*, Sydney: Federation Press.

Dovers, S. 2006, 'Precautionary policy assessment for sustainability' in Fisher, E., Jones, J. and von Schomberg, R. (eds), *The precautionary principle and public policy decision making*, Cheltenham: Edward Elgar.

Dovers, S. and Wild River, S. (eds) 2003, *Managing Australia's environment*, Sydney: Federation Press.

DRE (Department of Resources and Energy) 1983, *Water 2000: a perspective on Australia's water resources to the year 2000*, Canberra: AGPS.

Fisher, D. E. 2007, 'Delivering the National Water Initiative: the emergence of innovative legal doctrine' in Hussey, K. and Dovers, S. (eds), *Managing water for Australia: the social and institutional challenges*, Melbourne: CSIRO Publishing.

Frawley, K. 1994, 'Evolving visions: environmental management and nature conservation in Australia' in Dovers, S. (ed.), *Australian environmental history: essays and cases*, Melbourne: Oxford University Press.

Gowland, D. and Aiken, M. 2003, 'Privatisation — a history and survey of changes in organization structures, cultural and environmental profiles', *Australian Journal of Public Administration* 62: 43–56.

Hussey, K. and Dovers, S. 2007, *Managing water for Australia: the social and institutional challenges*, Melbourne: CSIRO Publishing.

Jackson, S. and Morrison, J. 2007, 'Indigenous perspectives in water management: reforms and implementation' in Hussey, K. and Dovers, S. (eds), *Managing water for Australia: the social and institutional challenges*, Melbourne: CSIRO Publishing.

Marsh, I. 2002, 'Governance in Australia: emerging issues and choices', *Australian Journal of Public Administration* 61: 3–9.

Marsh, I. and Yencken, D. 2004, *Into the future: the neglect of the long term in Australian politics*, Melbourne: Black Inc.

Powell, J. 2000, 'Snakes and cannons: water management and the geographical imagination in Australia' in Dovers, S. (ed.), *Environmental history and policy: still settling Australia*, Melbourne: Oxford University Press.

Proust, K., Dovers, S., Foran, B., Newell, B., Steffen, W. and Troy, P. 2007, *Climate, energy and water: accounting for the links*, Canberra: Land & Water Australia. ISBN 978-1921253348. www.lwa.gov.au/downloads/publications_pdf/ER071256.pdf

Richardson, B. and Bosselman, K. (eds) 1999, *Environmental justice and market mechanisms: key challenges for environmental law and policy*, London: Kluwer Law International.

Sheil, C. and Leak, B. 2000, *Water's fall: running the risks with economic rationalism*, Sydney: Pluto Press.

Smith, D. I. 1998, *Water in Australia: resources and management*, Melbourne: Oxford University Press.

UN (United Nations) 1992, 'Agenda 21: the programme of action from Rio', New York: UN.

Wanna, J. and Weller, P. 2003, 'Traditions of Australian governance', *Public Administration* 81: 63–94.

WCED (World Commission on Environment and Development) 1987, *Our common future*, Oxford: Oxford University Press.

Chapter 6

Sustainability in urban water futures

Geoff Syme

There is now a substantial literature defining and encouraging sustainable urban water management. Nevertheless, the responses of urban water utilities to changed demand and supply options have tended to focus on technological solutions, new sources and well-worn approaches to demand management. While there is increasing interest in water-sensitive urban design, whole-of-lifecycle economic consideration and the incorporation of externalities into pricing and cost-benefit analyses, there are significant areas of sustainability that have received scant attention. These neglected areas tend to relate to the difficulty in creating, as opposed to promoting, the concept of sustainability that includes social and cultural assessment, integrated response and key institutional issues in achieving adaptive learning.

From the community's perspective there are those who will be happy with modest modifications to the status quo. Recently, however, there have been attitude changes towards supporting stronger approaches to sustainable water-resources management. There is also a greater appreciation that holistic approaches will have to be taken as the issues associated with metropolitan growth and climate change have become more evident.

Thus it is realised by many in the community that there are important value judgments that will have to be addressed to establish whether the status quo should be maintained. Alternatively, if strategic social and cultural goals are to be achieved the community are willing to engage in examining what novel institutional structures should be seriously considered to attain them. Community engagement will need to be structured around the drivers of their decision-making and consequent water-supply preferences if procedurally just change is to be achieved.

Research has shown that these drivers include judgments on issues of fairness in allocation, acceptable risk and uncertainty, trust in both government and its agencies, and perceived wellbeing from alternative levels of service. Emotion is also a significant driver of community decision-making. These judgments are underpinned by perceptions of professional roles and knowledge and how they are incorporated in public discussion.

The above issues and the overall prospects and justification for change in decision-making in urban water management are discussed in terms of examples of community water culture in relation to alternative delivery systems and inter-regional transfer of water resources.

The important role that water plays in the urban environment is well documented. The amenity provided by traditional centralised water supplies overcame early water-borne diseases early in our history. Large sewerage systems led, with some isolated exceptions, to a comfortable 'out of sight and out of mind' mentality in regards to the inconvenient truths associated with waste disposal. Stormwater management was only an issue during heavy rainfall and if drains became blocked. Local wetlands and parks have been regarded as a metropolitan 'staple'. Economically, the amenity associated with proximity and access to water bodies in the landscape has been reflected in land and housing prices.

Thus, through its ready availability and largely unseen management, water, as with oil and electricity, has been largely taken for granted by city-dwellers. Recent challenges through drought and climate change, however, have left water planners, politicians and the public with difficult trade-off choices with regard to the costs associated with maintaining current levels of service, how to create acceptable demand management and the ongoing socio-political problems associated with a lesser level of security of supply and imposition of ongoing restrictions of differing degrees of severity.

The debate about levels of service and their ongoing contribution to community wellbeing and environmental sustainability is now on in earnest (for example, Larsen and Gujer 1997). Much of the sustainability-related argument in Australia is discussed in the context of the concept of integrated urban water management (Mitchell 2006). The resolution of this debate is urgent (for example, see Vlachos and Braga 2001). If historical expectations of the community in this regard are to be met, alternative sources of supply need to be identified with some alacrity. While decisions need to be made in a timely fashion, knee-jerk responses which are potentially unsustainable need to be avoided. The merits of 'quick fixes' such as the current penchant for desalination plants need careful analysis against alternatives. Some would say that technological 'fixes' such as desalination plants are unsustainable (Hurlimann 2007).

This situation reinforces the need for better long-term planning, as the limitations of our past efforts have been exposed. But to improve in this domain we also need to carefully examine the current assumptions upon which long-term demand predictions are made and our expectations of socially acceptable demand-management programs. This examination should, perhaps, occur at a very fundamental level. Currently, our explicit debates around acceptability of levels of service tend to focus around the relatively uninterrupted quantity of potable water provided and the aesthetics and risk associated with the system-level quality of that water. Furthermore, while holistic thinking is being promoted in the sustainability debate, levels of service seem to be discussed and investigated in a component context. That is, for example, required reliability is considered as one dimension, pressure as another, and health concerns and water aesthetics as two others. Regardless of our role in urban water management, the processing of each of these dimensions into an overall level of service is problematic for us all.

There is, of course, no reason to assume that the public expect that traditional levels of service or large-scale central delivery systems should be retained. The water-reform process and the increasingly evident community-based support for water conservation and sustainable development (Syme and Hatfield-Dodds 2007) indicate that the community may be highly supportive of responsible change in this area. Opponents of this view may tend to point to such issues as the rejection of potable supply from recycled water as an example of community conservatism in this regard. An equally plausible explanation may relate to the quality of the decision-making process and the crisis atmosphere in terms of lack of water and therefore confidence in the planning system.

Changes in water-supply delivery are often couched in terms of the need for urgent action to get more giga-litres in the dam. It is the contention of this paper that long-term planning may be better driven by examining the benefits to the urban community and how they can be delivered, rather than from the point of view of the amount of water in the dams. The question should be: Given the degree and range of benefits, how much water needs to be delivered and how can that best be achieved?

In short, the issues associated with the long-term planning of sustainable urban water supply should be addressed by starting with the holistic concept of wellbeing or the related concept of quality of life (Cummins *et al.* 2003; Pacione 2003) or the benefits to householders provided by active and passive use of water (Moran *et al.* 2004). Hoekstra *et al.* (2001) and Syme *et al.* (2008) point out that potentially the same volume of water can provide for multiple benefits. Potentially, therefore, demand management and setting levels of service is about using water most efficiently to retain these benefits. These will not only include the benefits at personal household levels but those which manifest themselves in the context of sustainability. These include the costs to the environment of augmenting current supply or re-allocating water from other users. As Syme and Nancarrow (2007) discuss, there appears to be a significant and strengthening aspiration in the urban community for sustainable urban-water supply and this goes beyond the water supplied to individual households. As has been foreshadowed (Syme and Hatfield-Dodds 2007) and will be elaborated in the following discussion, these wider sustainability concerns extend to institutional and decision-making process, trust and justice concerns.

Moving towards more sustainable urban water resource decision-making.

ARCWIS (1999) in their review of the social issues propose that the concept of levels of service be expanded from the traditional elemental approach to a broader perspective that includes issues related to the externalities of water supply as well as wider symbolic, ethical and aesthetic issues (see Table 6.1).

Table 6.1: Scoping Levels of Service (from ARCWIS, 1999)

	DIMENSION 1 What Comes in the Household's Door	DIMENSION 2 Costs of What Comes in the Household's Door	DIMENSION 3 Symbolic, Aesthetic, Moral or Ethical Issues
Quality of Service Delivery	1. time taken to answer phones 2. response time to complaints 3. design of bill 4. etc.	5. loss of environment 6. risk of dam failure 7. water treatment 8. recreation	9. equity 10. who pays 11. definition of Community Service Obligations 12. regional differences in levels of service
Quality of Product	13. Pricing 14. Drinking water quality 15. Water pressure 16. Reliability of supply 17. Etc.	18. Waterways quality 19. Risks in implementation 20. Confidence in institutional arrangements	21. Appropriate degree of recycling 22. Support of urban amenity

While this is a significant broadening of issues for consideration by urban utilities there is no guidance given within the report as to how these dimensions should be integrated to arrive at an overall judgment. Perhaps more importantly, key process issues associated with implementing change without equity issues becoming a significant barrier is not specifically addressed. Some questions that may arise in regard to equity issues are outlined in what follows.

Given that new technology may make it possible for individual households to treat and recycle their own wastewater, how should we manage if a large proportion of households decide to opt out of a central system? Should that system be retained and, if so, who should pay for the upkeep of it? These equity issues in relation to change are likely to be much easier to resolve or accommodate in new suburbs where a uniform approach can be taken in a locality and, on purchasing the house, the householder can voluntarily opt in to an alternative system. This is probably why most reform is currently occurring in new suburbs rather than in established areas. The question is, given the traditional behaviour of the urban water industry is it competent or appropriately constructed to manage significant change in the delivery of urban water from the viewpoint of equity and related social constructs?

The discussion around change is much more than the macro issues such as whether the water industry should be privatised. It includes more subtle judgments in regard to who should manage the urban water cycle, why and at what level. Theoretically, if home treatment were to become the preferred option much of the water cycle could be managed by small-to-medium enterprises on a regular inspection basis. In many ways, we could be returning to the days of the 'night cart'. This is a return to social organisation and practice current more than a century ago — a long way from centralised management, which currently enjoys the confidence of the public.

The problems associated with defining sustainability in urban water management are challenging enough, but achieving it with integrated macro-, meso- and micro-systems may prove to be a very high hurdle. Cities change incrementally and water-management reform occurs gradually, driven by pressures for more urban development. Even when integrated urban water-resource management is mooted by state government, such as in New South Wales, progress is in the form of pilot studies (Anderson and Iyaduri 2003). In this situation there may be no formula that that can provide a silver bullet for reasonably uniform change across entire cities. This add-on approach will need a diversity of tactics and strategies for water management across differing areas of cities.

Perhaps the best contemporary exemplar of the problems facing reform in the urban water industry can be seen in the challenges occurring in managing urban stormwater and urban lakes and streams. In Western Australia, despite several workshops and studies over 30 years (including the preparation of a multi-stakeholder based White Paper for government after two years' discussion and a universal agreement on a set of sustainability principles) reform remains elusive. Funding and implementation issues still cannot be agreed between state and local government. In slightly different forms this problem occurs throughout Australia. Perhaps this is understandable given the complex nature of stormwater, but if the other components of the water-management cycle become more fragmented similar difficulties in governance and management may arise.

So how can we create change from the relatively simple centralised 'catchment to the sea' systems to manage water in a whole-of-water-cycle sense that which may require smaller infrastructure and more-localised management? Do we want to and do we have to? Our response will probably be governed largely by the wider socio-political system in which urban water is currently managed, in particular state government politics and demands. Issues such as the need to provide financial returns to the government, the sensitivity of voting intention and the poor definition of the public-good components of urban water supply will all tend to hinder reform beyond tariff structures and the introduction of technological 'fixes' such as desalination plants. Neither of these will lead to cultural change in regard to urban water and its provision. In many respects it is the urban development industry that is leading reform on a locality-by-locality level and the market is beginning to reward innovation, although it is far from certain as to whether it is the water innovation that governs purchasing behaviour. Mitchell (2006), for example, reviews a significant number of infrastructure developments that are occurring in the Integrated Urban Water Management paradigm using total-water-cycle management concepts. Despite their equal importance, examples of the enabling of water reform in this regard from the viewpoint of social and decision-making processes are remarkably fewer.

Can social research help reform?

Given this slow-moving system that largely operates on political hunch and market research it would seem that detailed knowledge of what futures the community sees as sustainable and what trade offs the community is prepared to consider would be of great value. Thirty years of experience with community-based research have led to an observation that, given a facilitative environment, the community is often prepared to make choices which are decidedly more innovative than those currently being made on their behalf. For example, in the 1980s a carefully constructed longitudinal field-based study of the latitudes of tolerance (and intolerance) of a variety of possible restrictions policies in Perth was conducted (Nancarrow, Kaercher and Po 2002). The findings were very definitive. By far the preferred policy was for the then Level 1 restrictions to become permanent. On presentation of the results to the then Minister for Water Resources, these findings were dismissed. We were later told that it had been surmised at a political level that a state government had lost an election in the 1960s because of water restrictions. Of course it is hard to define whether and how such a 'political urban myth' affected the decidedly limp response to the social research. Nevertheless, it would seem that, when in doubt, the political environment will tend to prevail over service. The traditional engineering decision-making culture (Spearritt 2007) would tend to underscore this tendency.

One area where reform can be assisted by social science is that of ex-ante evaluation of innovative technologies. This has been the purpose of a long-term 'water cultures' program (Leviston, Porter and Nancarrow 2006) in which a generalised community decision-making model has been derived from examination of a number of innovations in several Australian cities (for example, high-rise water-recycling schemes, managed aquifer recharge, suburban recycling schemes and so on). A number of replicable elements to community-based decision-making have been found (see Figure 6.1). The models have gradually been refined and now can explain a very high proportion of the variability in acceptance (upwards of 80 per cent).

Figure 6.1: A structural equation model of the acceptability of novel water-supply systems (Leviston, Porter and Nancarrow, 2006)

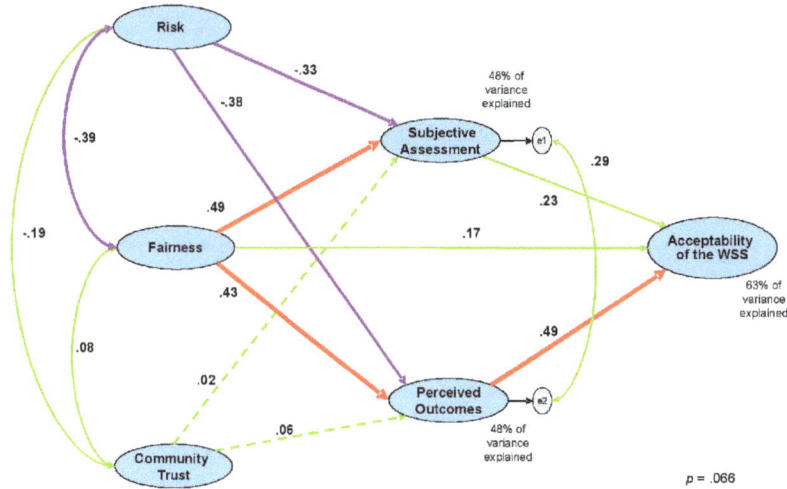

While it is not the purpose of this chapter to comment on the variables in detail, three important determinants can be related to the institutional context for achieving change. Institutions in this context relate not only to the formal governmental organisations or utilities but also the more informal groupings and cultural values embedded in the community. People tend to see acceptable sustainable reform in terms of institutional behaviour as well as the familiar triple bottom line. These institutionally related considerations are perceptions of acceptable risk, perceived fairness of the new system to all consumers and trust in the water utility and its decision-making processes. To achieve reform it is important, therefore, to spend some time on these variables to foster a process that can incorporate ongoing innovation. While there is obviously a need to understand the dynamics of risk, trust and fairness and how to create them, particular attention should be given to the concept of fairness as this has a strong shorter- as well as a longer-term component.

The overall judgment of fairness consists of three components (Tyler and Blader 2000; van den Bos and Lind 2002): distributive justice, relating to the proportion of benefits and costs that differing groups in society; procedural justice — whether the community feels that it has had an adequate opportunity within the decision-making process to make comment (for example, did it have a 'voice' or an opportunity to have an influence?); and 'interactive' justice, which is sometimes considered as a component of procedural justice. Simply stated, interactive justice reflects whether people feel as though they have been treated with respect and valued as individuals during the decision-making process.

Overall judgments of justice have two components, cognitive and affective. That is, fairness judgments involve logical thinking in relation to the issue at hand. These judgments also have a strong and significant emotive part. Emotion is important, as it is exhibited as a consequence of reactions to the perceived justice of treatment that has been given to the individual or his community (de Cremer and van den Bos 2007). For example, one can easily imagine that a perceived lack of procedural or interactive justice could be important in engaging the emotional side of fairness judgments. These feelings may also lead to an enduring sense of outrage and protest, particularly in the areas where risk is involved.

Thus, those who hope that the decision-making processes will remain rational and that emotion will be avoided would seem to be doomed to disappointment. Emotion is a basic part of human nature and an important component of democracy. The important point for this discussion is that getting our public processes 'right' will be a powerful driver of reform, especially as it is so hard to achieve within our current socio-political environment. The community can only participate and lead in reform if effective decision-making processes are put in place.

The cogency of this argument is also highly evident in a behavioural model developed by Leviston *et al.* (2006) for predicting actual individual consumption of recycled wastewater for potable purposes. This model (see Figure 6.2), the variables for which were derived from a community-based experimental study, like the 'water culture' model underlined the significance of the above discussion. As for the acceptance of new water-delivery systems, generally key variables for willingness to drink such water were trust in the agency, emotion (disgust or the 'yuck' factor) and perceptions of risk. The two other variables of significance were also of interest in predicting actual consumption of the water and, potentially, for conducting the public decision-making process. These were subjective norm, or the opinions of other significant others about the issue, and perceived control over the necessity to drink recycled water. Subjective norm tends to indicate that there will be conferring between individuals about the subject that is likely to lead to the desire for public debate.

Perceived control would seem to indicate that if people feel that they have been forced to drink recycled water by poor planning or lack of procedurally just decision-making processes, they will try to avoid it. Both the subjective norm and perceived control variables underscore the need for good participative processes.

Figure 6.2: A structural equation model of intended behaviour to drink recycled water (Leviston *et al*. 2006)

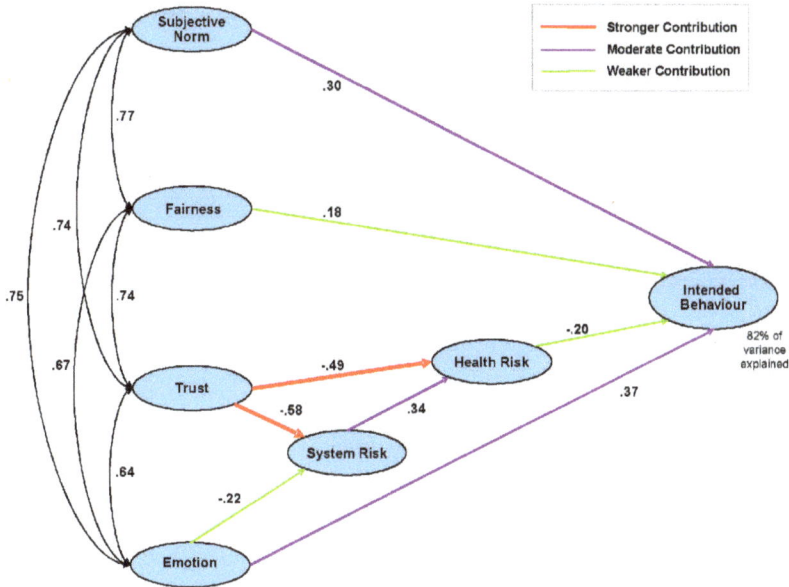

In short, detailed, in-depth and preferably longitudinal social science research can provide a map of good process which can deliver on the fairness issues. Because the models show significant interrelationships between the achievement of fairness, the development of trust and more positive views of perceived risk, this fairness-based decision-making map becomes a central tool for planning and implementing productive reform.

Relating social science to adaptive learning and sustainability

Initially, sustainability was often considered in the context of evaluating plans and practice through the triple bottom line of environmental, economic and social analyses. Apart from the community, there is an increasing professional acknowledgement that a key element of sustainability is institutional (for example, Hedelin 2007). That is, it is recognised that the quality of the decision-making process is a highly significant precursor to implementing sustainable practice. Institutions are not simply construed as formal government organisations but more as the interfaces of various sectors of the community and government to obtain outcomes which reflect the long-term aspirations of the community for the benefits it wishes to attain from its water resources (Syme *et al*. 2008). As described earlier, one of the key elements of high-quality decision-making in this context is the achievement of procedural justice in the involvement the community has with the water utility.

Procedural justice has a number of elements (Lawrence *et al.* 1997) and these support acceptance of the decision-making process and its outcomes (Tyler and Blader 2000). While most of this social-justice research has been conducted in traditional social service areas, it is relevant to urban water management and other natural-resource management issues. For example, in a long-term planning exercise for creating strategic plans for the use of wastewater with the West Australian Water Corporation, perceptions that the public-involvement process was procedurally just led to greater commitment for involvement in future decision-making activities (Syme and Nancarrow 2002).

In procedurally just processes because there is voice (or an open expression of views), which enables information exchange, and decision-making can include local knowledge that can enhance the possibilities for adaptive learning. Adaptive learning is an important underpinning of sustainability in that increasing knowledge can contribute to the wisdom of decision-making and the avoidance of 'surprises'. While the concept of adaptive learning is appealing, it has not really occurred in many places. But if adaptation is to evolve in an atmosphere of continuous improvement, it is vital that urban utilities grasp the opportunities provided by procedural-justice research and practice in the sustainability context.

The recent debate on whether to pipe water to Perth from the large South West Yarragadee aquifer, some 300 kilometres from the metropolitan area, provides a good example of the widening context within which urban utilities will be evaluated. This transfer would assist in lessening the probability of the imposition of severe restrictions. The early social research on this issue indicated concern by both South West and Perth residents that there was too little knowledge of the aquifer; fairness issues relating to the concept of 'reasonable regional needs' (that is, was it fair to take the water from that community?); and finally the doubt that the Water Corporation would reverse its decision if data showed that it was not sustainable for the aquifer. These issues continued to drive the community debate as the hydrological modelling became more detailed and the social and economic analysis increased. Finally, it became evident that the limits to the usefulness of modelling had been reached and that only monitoring the effects of the abstraction of water would validate the decision. The Independent Sustainability Panel established by the State Government, therefore, in its acceptance of the proposal put in place a series of public-accountability and involvement processes designed to ensure that adaptive learning through procedurally just processes could occur (Sustainability Panel 2007).

The issue for the community then became whether they could trust the utility or the regulators to implement the recommendations effectively. Comparisons were made between the recommended adaptive-learning approach and the widely publicised deterioration in the Gnangara mound aquifer (a groundwater mound also supplying Perth), where breaches in extraction rules had been publicised. Thus the issue of trust was paramount for the Water Corporation and for the community when assessing new developments.

Given the growing discussion of reallocation of water from rural to urban communities, this example is likely to recur. Thus, the water industry has been moved by circumstance and the community to become a much more visible player in the overall movement towards adaptive learning and sustainability. It is no longer sufficient for urban utilities to think on a project-by-project basis and to deliver levels of service that satisfy only the utilitarian aspects of water supply. Increasingly, issues of urban and environmental amenity and the integrity and quality of decision-making are becoming central concerns for the community. Longitudinal social research evaluating the maintenance of procedural justice and providing a basis of understanding the generation of trust while incorporating the concept of acceptable risk can provide a major contribution to the achievement of sustainability and improved decision-making. This contribution is currently patently lacking.

Can we usefully study institutions to help achieve sustainability?

Saleth and Dinar (2000) have identified a series of endogenous and exogenous factors that will tend to promote water institutional changes. Exogenous variables of particular relevance to Australia are water scarcity, performance deterioration, financial non-viability and the emergence of technological progress. These conditions, the authors suggest, will create a need for more adaptive institutions with lower transaction costs and a pro-reform environment. All these conditions for institutional reform are highly evident in Australia. Establishing efficient and effective formal and informal institutions would therefore seem to be logical if sustainability is to be achieved to a greater extent than it currently is in the Australian context. Institutional research within an integrated water-management framework should therefore have a high priority. Nevertheless, holistic research in this area has been scarce.

While specific alternative institutional or management structures were part of the scenarios included in the ARCWIS analysis of the community's willingness to accept water-supply innovation, this was not detailed enough to determine any generic findings. Are there any general rules in relation to institutional structure and functions that can assist in the implementation of procedurally just decision processes? This type of question can be answered in the community context in urban water-resources management. Ostrom (1990) has provided extensive work on understanding the principles and criteria for successful institutional arrangements for management of common property resources for small and largely rural communities. The same level of analysis will be required for alternative forms of governance for urban water. There is a need for social and organisational research that focuses on the generics of institutional structure, decision-making processes and implementation responsibilities.

This research should be conducted with water-resource decision-makers who can provide ideas for alternative institutional arrangements and functions that could support alternative but coherent arrays of local, meso and whole-of-system management. These could be considered in the light of each element of the water cycle. It should deal directly with issues such as the appropriate communication and responsibility networks, the role of legislation, public versus private water-supply responsibilities, transaction costs, and so on. For sustainability purposes these should be interpreted in the light of social, economic and environmental analysis. As the previous discussion in this chapter indicates, the social analysis and preferred network identification and the overall sustainability goals of whole-of-water-cycle management will also need to incorporate public perspectives to create a management framework that will create ongoing flexibility for change.

The investigations need to be inclusive of differing breadth of decision-making from macro-allocation decisions (for example, inter-regional transfers from rural to urban regions) to more micro decisions (alternative localised stormwater management). The need for social and institutional bottom lines has been shown to be imperative at both levels. Both levels also incorporate much more than the current rational decision-making paradigm. There is a need for the incorporation of structures and processes that can cope with emotions and ethics of fairness and organising frameworks such as trust and accountability. Perceived and acceptable risk is also particularly important when water quality-related issues are discussed. These variables can be measured in an ex-ante and formative manner (Kasemir *et al.* 2003) to derive an adaptive-learning approach to urban water reform.

Conclusions

The traditional approach to sustainability in urban water provision has been to provide a series of indicators. Social, institutional and decision-making issues have been discussed in general terms (Kenway, Howe and Maheepala 2007) but are generally included as a vague backdrop unless quantitative demographic information is obtainable. Thus they are there but are not seen as principal drivers of change. But if change is to occur, it is these dynamic social things that can drive it: ethics, values, attitudes and behaviour at individual, organisational, community and political levels.

If we are to advance towards sustainable water futures it is important to free up those things which are currently holding it back. The social, institutional and cultural variables thus become of paramount importance. These must be underpinned by processes that are seen to be procedurally and interactively just. Once these processes are in place, economic and environmental investigations come into play, as do the more conventional social-sustainability indicators. It is analysis of alternative scenarios from these perspectives that provides important criteria against which alternative futures can be described.

Finally, it is worth observing that this chapter does not provide the right answer to which way the urban water industry should go forward. Centralised systems have provided reliable and healthy water supply in the past, albeit by providing three largely separate systems. But such infrastructure has been facing large challenges in Australia in recent years because of drought and increasing problems with water quality. On the other hand, micro-systems may provide for sustainability and whole-of-cycle management at a micro-level but face management and regulation problems. Meso-scale developments, while being easier to manage than micro-systems, could face issues of social acceptance, especially if they are in the form of neighbourhood wastewater treatment plants.

In reality, if there is to be a change from large centralised systems and there is a consistent introduction of meso- and micro-systems, for whatever reason, there will be major challenges for the urban water industry, both socially and institutionally. Most particularly, widening the scope from rational engineering and economic decision-making to incorporating community values and emotions into procedurally just and iterative decision-making processes will be the greatest challenge of all. Without integration of these components, sustainable urban water systems are unlikely to be achieved.

References

Anderson, J. and Iyaduri, R. 2003, 'Integrated urban water planning: big picture planning is good for the wallet and the environment', *Water Science and Technology* 47(7–8):19–23.

ARCWIS 1999, *The Social Basis for Urban Water Provision in the 21st Century*, CSIRO Urban Water Program, CSIRO Land and Water, Perth.

Cummins, R. A., Eckersley, R., Pallant, J., Van Vugt, J. and Misajon, R. 2003, 'Developing a national index of subjective wellbeing: The Australian Unity Wellbeing Index', *Social Indicators Research* 64: 159–90.

De Cremer, D. and van den Bos, K. 2007, 'Justice and feelings: Towards a new era in social justice research', *Social Justice Research*, 20 (1): 1–9.

Hedelin, B. 2007, 'Criteria for the assessment of sustainable water management', *Environmental Management* 39: 151–63.

Hoekstra, A. Y., Savenije, H. H. G. and Chapagain, A. K. 2001, 'An integrated approach towards assessing the value of water: A case study on the Zambesi basin', *Integrated Assessment* 2: 199–208.

Hurlimann, A. 2007, 'Time for a Water Re-"Vision"', *Australasian Journal of Environmental Management* 14: 14–21.

Jenerette, G. D. and Larsen, L. 2006, 'A global perspective on changing sustainable urban water supplies', *Global and Planetary Change* 50: 202–11.

Kasemir, B., Jaeger, J., Jaeger, C. and Gardner, M. (eds), *Public Participation in Sustainability Science,* Cambridge University Press, Cambridge.

Kenway, S, Howe, C. and Maheepala, S. 2007, *Triple Bottom Line Reporting of Sustainable Water Utility Performance,* AwwaRF, in press.

Larsen, T. A. and Gujer, W. 1997, 'The concept of sustainable water management', *Water Science and Technology* 35 (9): 3–10.

Lawrence, R. L., Daniels, S. E. and Stankey, G. H. 1997, 'Procedural justice and public involvement in natural resource decision making', *Society and Natural Resources* 10: 577–89.

Leviston, Z., Nancarrow, B. E., Tucker, D. I. and Porter, N. B. 2006, 'Predicting community behaviour: Indirect potable reuse of wastewater through Managed Aquifer Recharge', CSIRO Land and Water Science Report 29/06: Perth.

Leviston, Z., Porter, N. B. and Nancarrow, B. E. 2006, *Interpreting householder preferences to evaluate water supply systems,* Stage 3, Water for a Healthy Country National Research Flagship, CSIRO Land and Water: Perth.

Mitchell, V. G. 2006, 'Applying integrated urban water management concepts: A review of Australian experience', *Environmental Management* 37(5): 589–605.

Moran, C. J., Syme, G., Hatfield-Dodds, S., Porter, N., Kington, E. and Bates, L. 2004, 'On Defining and Measuring the Benefits from Water', paper presented at 2nd IWA Leading Edge Conference on Sustainability—Sustainability in Water Limited Environment, November 2004, Sydney.

Nancarrow, B. E., Kaercher, J. and Po, M. 2002, 'Community Attitudes to Water Restrictions Policies and Alternative Sources. A Longitudinal Analysis, 1988–2002', CSIRO Land and Water Consultancy Report: Perth.

Ostrom, E. 1990, *Governing the Commons: the evolution of institutions for collective action,* Cambridge University Press: Cambridge.

Pacione, M. 2003, 'Urban environmental quality and human wellbeing — a social geographic perspective', *Landscape and Urban Planning* 65:19–30.

Saleth, R. M. and Dinar, A. 2000, 'Institutional changes in global water sector: trends, patterns and implications', *Water Policy* 2: 175–99.

Spearritt, P. 2007, URBAN WATER: Crisis and response. This symposium.

Sustainability Panel 2007, *Sustainability Assessment of the South West Yarragadee Water Supply Development*, Department of Premier and Cabinet, Government of Western Australia: Perth.

Syme, G. J. and Hatfield-Dodds, S. 2007, 'Values attached to Water and the potential for Communication and Institutions that Encourage Community and Government Partnerships to Achieve Implementation of Water Reform' in *Implementing Water Reform in Australia*, CSIRO Press: Collingwood.

Syme, G. J. and Nancarrow, B. E. 2002, 'Evaluation of public involvement programs: Measuring justice and process criteria', *Water* 29: pp.18–24.

Syme, G. J. and Nancarrow, B. E. 2005, 'Sharing the Pain of Water Reallocation: Creating Consent by Taking Fairness and Justice Seriously' in D. Cryle and J. Hillier (eds), *Consent and Consensus, Politics, Media and Governance in Twentieth Century Australia*, API Network, Curtin University of Technology, Perth, Australia: 371–87.

Syme, G. J. and Nancarrow, B. E. 2008, 'Changing Attitudes to Urban Water Use and Consumption' in P. W. Newton (ed.), *Transitions: Pathways Towards Sustainable Urban Development in Australia*, CSIRO Publishing, Collingwood. In press.

Syme, G. J., Porter, N. B., Goeft, U. and Kington, E. A. 2008, 'Integrating social wellbeing into assessments of water policy: Meeting the challenge for decision makers', *Water Policy*. In press.

Tyler, T. R. and Blader, S. L. 2000, *Cooperation in Groups: Procedural Justice, Social Identity and Behavioral Engagement*, Psychology Press: Philadelphia.

Van den Bos, K. and Lind, E. A. 2002, 'Uncertainty management by means of fairness judgments' in M. P. Zanna (ed.), *Advances in Experimental Social Psychology* 34, Academic Press: San Diego.

Vlachos, P. E. and Braga, P. B. 2001, 'The challenge of urban water management' in C. Maksimovic and J. A. Tejeda-Guibert (eds), *Frontiers in Urban Water Management, Deadlock or Hope?*, IWA Publishing, London: 1–34.

Chapter 7

Exploiting the unspeakable: Third-party access to sewage and public-sector sewerage infrastructure

Janice Gray and Alex Gardner

Water has the capacity to capture the imagination, particularly the Australian imagination. Historically, it has loomed large in Australian literature, film and art, both overtly and suggestively. Sometimes it has been a site of celebration, sometimes a site of pain and darkness. The breadth of water imagery is vast. A cursory reflection brings to mind Patterson's *Clancy of the Overflow*,[1] Slessor's *Five Bells*, Winton's *Cloudstreet,* Drewe's *The Drowner* and *The Bodysurfers*, as well as Streeton's painting *Sunlight Sweet, Coogee*, and Done's prolific collection of harbour and beach paintings, for example.[2] The subject of water in these representations is commonly fresh, clean water. Often beginning in pure mountain springs, it gushes down rivers displaying strong, untamed qualities. At other times, the representations are of the salty seawater that provides pleasure for a sun-loving beach culture or, alternatively, they are images of freshwater in life-sustaining rural rivers — part of a lost pastoral era. In bolder and more sinister representations, water may take life. Waves and tsunamis may cause death. However, rarely have art, literature and film taken as their subject water that is polluted by waste, particularly human bodily waste. That subject is left largely to science. The film *Kenny* was an exception, relying on 'toilet humour' and focusing, as it did, on the provision of waste removal. Sewage, outside of the sciences, is largely what might be called 'unspoken' water.

This chapter suggests that 'unspoken water' is rapidly becoming 'spoken' water. Sewage is in the process of being 're-imagined'.[3] A paradigm shift is taking place in the community's attitude towards wastewater and sewage. Water which was once seen as a nuisance is now being recast as a valuable resource. New technology is being developed to treat and use this resource for industrial, agricultural and, even, environmental and drinking purposes (although, as Spearritt observes in Chapter 2, the 'yuck' factor posed difficulties in relation to the water referendum in Toowoomba). The recycling of human wastewater is being reinvented as both an environmental and commercial opportunity that can be facilitated by giving 'third-party' access to established public-sector sewerage infrastructure and — importantly — to the sewage! The realisation of the resource value of wastewater has spawned a new industry — sewer mining!

But there is a problem — a legal problem. Historically, large public utilities have controlled urban sewerage infrastructure and they have not been enamoured with new private-sector operators entering the field as competitive service providers by means of access to their infrastructure. With some justification, they see the infrastructure and the wastewater resources as their property. Yet, they may not possess the technology or capital to exploit the new water resources in the timely way that the paradigm shift in attitudes suggests is opportune. Conversely, the new private-sector entities may not be able to operate without obtaining access to the established public-sector infrastructure, either to provide a water supply or sewerage service, or to access the wastewater resource with a view to converting it into a saleable product. So, we confront the question of how to provide new third-party (and usually private-sector) access to old public-sector infrastructure in order to make better use of these valuable water resources. We also confront the associated issues of how to maintain health standards and societal protections in the face of private-sector involvement in the supply of fundamental life services.

Providing for private-sector access to wastewater infrastructure and facilitating private-sector wastewater services will require sophisticated levels of science and technology. It will also require the development of an appropriate legislative framework to regulate the private-sector access and services. Regulatory wastewater regimes will need to operate in tandem with the broader legal framework for water-services provision, including the economic regulatory bodies at both the State and Commonwealth levels.

This chapter seeks to highlight where opportunities may exist for third-party access to public infrastructure and, accordingly, it sketches the present institutional frameworks for water and wastewater management throughout Australia. It then briefly reviews the international and national experiences of third-party access in the water sector and discusses the relevant Commonwealth legal framework for facilitating third-party access to monopoly service infrastructure, contained in Part IIIA of the *Trade Practices Act* 1974 (Cth) *(TPA)*. In particular, we consider the catalytic role of Services Sydney Pty Limited, the private company seeking access under that Act to the sewerage network operated by the Sydney Water Corporation (Sydney Water), a public water utility. The company's application for a 'declaration' that Part IIIA *covers* the sewerage service operated by Sydney Water demonstrates the access regime in operation and, effectively, gives us a 'test case' of the concept.

The chapter then reviews the *Water Industry Competition Act* 2006 (NSW) as an example of state-based industry-specific legislation incorporating a third-party access regime. This legislation may become a model for other states (such as Western Australia) as they seek to increase opportunities for competition in the water sector. The final section of the chapter attempts to tease out some points of interest and potential concerns associated with third-party access regimes. It considers issues of water-service coverage declarations, access and resource pricing, building customer trust in new third-party service providers, step-in rights for operators of last resort when a service provider fails, 'property' in the water resource, and consumer-protection provisions.

We acknowledge at the outset that the sorts of issues discussed here in the context of wastewater may similarly arise in relation to drainage water and drainage infrastructure. However, this chapter reviews only the situation in relation to wastewater.

Current institutional frameworks for water and wastewater

In order better to appreciate where opportunities for third-party access may potentially arise, a brief survey of the institutions currently responsible for the supply of water and wastewater services throughout Australia is offered. They are discussed on a state-by-state basis.

In Queensland, 125 local governments have been responsible for the supply of water in all urban areas, while in a number of regions some bulk water supply is provided by corporatised entities of the state government. Changes to this structure are likely in response to the Queensland Water Commission's (QWC's) Final recommendations, which were designed to allow third parties access to natural monopoly segments in the supply chain. The QWC Report, among other things, recommended that the 25 water entities in Southeast Queensland be collapsed into nine and that the State take control of wastewater in the region (QWC *Final Report*: May 2007; QWC *Fact Sheet* 1: 1).

In Victoria, four main metropolitan water providers operate, with Melbourne Water acting as the wholesaler, providing bulk supplies and wastewater services to three large distribution and retail service providers (AGD of PMC August 2006: 2). The distribution and retail businesses buy in water and then distribute it to their customers. Later, they collect the wastewater, bill customers and send the wastewater to Melbourne Water, where it undergoes treatment. No direct competition occurs between the three distribution and retail businesses because they all operate within different geographic boundaries. In Victoria, there are also 15 non-metropolitan urban service providers operating outside of Melbourne. They service country towns. Another five regional water authorities provide water, mostly for irrigation purposes (AGD of PMC August 2006: 2).

In Western Australia, South Australia and the Northern Territory, large state-wide utilities operate in both rural and urban areas. In Western Australia, the Water Corporation provides most of the water and wastewater services, as well as infrastructure building.[4] It contracts out some of these tasks to the private sector (through a tendering process) while it performs other tasks itself.[5] In 2007, the WA Economic Regulation Authority received a brief to explore the issue of increased competition in the water and wastewater sector. One aspect requiring particular attention was third-party access to water and wastewater infrastructure (ERA Issues July 2007: *i*).

In South Australia, SA Water operates as a wholly government-owned entity responsible for the provision of water and wastewater services throughout the state. It also contracts out many of its responsibilities to private operators. One such operator is United Water,[6] the sole and private retail provider of urban water services for Adelaide, to whom a $15\frac{1}{2}$-year contract was awarded in 1995. The contract covers the management of water and wastewater-treatment plants, water and wastewater mains, billing and the operation of call centres.

In Tasmania, retail supply and wastewater services are provided by local councils in regional areas, while in Hobart eight local councils jointly own Hobart Water, which services all the councils in the Hobart area.[7] In the Australian Capital Territory, a joint venture between the Australian Capital Territory Government and a private operator formed ActewAGL, which manages water and wastewater services (AGD of PMC August 2006: 2).

In New South Wales, the bulk water supply for Sydney, the Illawarra and the Blue Mountains is provided by the Sydney Catchment Authority, with retail services being supplied by Sydney Water, a state-owned corporation. The Sydney Catchment Authority was established as a result of the Independent Inquiry into Sydney Water's water-supply function, conducted by Commissioner Peter McClellan, in the wake of the 1998 Cryptosporidium and Giardia scare.[8]

Sydney Water provides water (including some recycled water), sewerage services and limited stormwater drainage services to Sydney, the Illawarra and the Blue Mountains. In the Newcastle area, Hunter Water, another state-owned corporation, provides both bulk and retail water-supply services. Local councils provide urban water services in all other towns and cities.[9] The State Government also manages separate bulk water supplies in various regions throughout New South Wales.[10] Overall 120 different water-service providers operate in New South Wales (AGD of PMC August 2006: 2).

We surmise that there would be many opportunities for third-party access to major water infrastructure as a means of supplementing the existing provision of water and wastewater services.

Finally, in all Australian states the primary legislation that facilitates the key functions and obligations of the public water utilities is supported by other legislation and regulations including those relating to state-based economic regulators, standards of water quality and use, local government, protection of the environment, public health and consumer protection. This suite of legislation would also apply to any operators who enter the water-services industry via third-party access to the monopoly infrastructure.

Definition of third-party access

As the name suggests, third-party access involves a party other than the incumbent service provider and infrastructure owner gaining access to and using excess capacity in the natural monopoly infrastructure of the incumbent (AGD of PMC August 2006: 13).[11] The purpose of the access is to promote competition in markets upstream or downstream of the relevant infrastructure.[12] Of course, the service provider could contract with the third-party access seeker to grant that access, but the provider may be unlikely to do so where the third-party proposes to compete with the service provider in the upstream or downstream markets. Certainly, the infrastructure owner possesses the unilateral capacity at general law to refuse access. The purpose of the statutory third-party access regimes is to facilitate such access and to provide a framework for an independent agency to arbitrate the terms of third-party access in those cases where the incumbent and the third party or parties cannot agree the terms of access. The presence of the statutory access regime would, no doubt, set the framework for any private access arrangements and may even be preferable for the incumbent service provider because it can provide procedural certainty.

As a result of the infrastructure access, the new entrant would be in the position to target new customers or the incumbent's existing customers and provide them with an established service, such as water supply or sewage removal. Alternatively, the third party could supply a new service as a result of gaining access. An example of the latter would be a third party supplying a water-recycling service as a result of gaining access to the incumbent's wastewater system[13] (AGD of PMC August 2006: 13). This has become known as 'sewer mining'.

Relationship between third-party access and sewer mining

Sewer mining is defined as the process of tapping into a sewer (before or after it reaches the sewage-treatment plant) and extracting the sewage so that it can be treated in a separate treatment facility and put to another use as recycled water (Sydney Water, *How to Establish a Sewer Mining Operation*, May 2006).[14] Although the production and supply of recycled water could be undertaken by the sewerage service provider, sewer mining is normally associated with third-party access by persons who either use the recycled water for their own purposes or supply it to others. In that sense, sewer miners are a subset of a wider class, known as third-party access seekers. Sewer miners engage in sewer mining by virtue of a contractual agreement with the public water infrastructure owner (for example, Sydney Water).

To date, sewer miners have established sewer-mining operations either over or adjacent to the sewer main on their own land. They have recycled the sewage at that venue and then piped the recycled water to the point of usage.[15] Several schemes of this nature are located on golf courses, where there is a demand for the product and enough space to build sewer-mining operations sufficiently distant from residents.[16] Nevertheless, sewer mining, particularly on a large scale, is still in its infancy despite the sewage farms begun at Botany Bay in the 1960s and the Werribbee Treatment Plant, noted by Dingle in Chapter 1.

Sewer mining does not involve third parties using the infrastructure owner's sewerage pipes to provide a competing sewage-removal service by transporting wastewater. Hence, a sewer miner would not contract directly with the incumbent's customers for the provision of a sewerage service, whereas the third-party access seeker who becomes a competitive service provider will do so. Accordingly, the third-party sewerage service competitor will be responsible for extracting a designated volume of sewage from the sewerage network commensurate with the input of the customers with whom the third-party has contracted to supply sewerage services. The extraction proportion is likely to be calculated on the basis of the third-party access seeker's proportion of all sewage at a given time or alternatively on the basis of an agreed volume, extracted at a fixed rate (*Sydney Water Network Access Agreement*, August 2007, cl 8.1).[17] Either way, the alternative sewerage-service provider may not vary, at will, the amount of sewage taken from the system. Further, the provider will be responsible for treating and disposing of its sewage share according to approved standards. Technically, third-party access seekers could provide a sewage-removal service *without* also engaging in recycling. However, we suspect that the potential profits of water recycling would be a substantial incentive for third-party access seekers endeavouring to gain access to provide a competing sewerage service.

There may also be other attractions to becoming a competing sewerage-service provider. Although the infrastructure owner must be able to maintain adequate flows in the sewerage system for both the competing service provider and the sewer miner, the amount of sewage available to the sewer miner is more variable. Minimum operational flow requirements, the diurnal flow pattern of the system and existing commitments to the extraction of sewage up or down stream will determine how much sewage is available for the sewer miner to extract.[18] The available sewage will then be allocated on a first-come-first-served basis, giving sewer miners some security of resource access to their operation (Sydney Water, *How to Establish a Sewer Mining Operation*: August 2007: 1). However, Sydney Water offers no guarantee of the quantity or quality of the sewage available for sewer mining.

Accordingly, in the context of the wastewater industry, a third-party access seeker may compete with the incumbent service provider in the upstream market (sewage removal) or the downstream markets (sewage treatment and supply of recycled water). In the Sydney context, the sewer miner operates in the downstream markets and has not been seen as a direct competitor of the incumbent service provider, Sydney Water. This is probably because the supply of recycled water is a relatively new and small market and the sewer miners have assisted Sydney Water by saving sewage-treatment costs and supplying recycled water to displace demand for treated drinking water. However, the company, Services Sydney, sought third-party access to compete with Sydney Water in the established upstream market for sewage removal as well as the downstream markets of sewage treatment and, eventually, the supply of recycled water. This evoked a very different response from Sydney Water, perhaps because the proposal threatens its control of the newly perceived resource — sewage.

Experiences of third-party access regimes for water

Third-party access has been available in network-based utility industries such as telecommunications, gas and electricity for some time[19] but the experience of third-party access in the water industry is somewhat limited.[20] In England, the *Water Act* 2003 seeks to extend competition, with s 39(2B) specifically seeking to protect consumers by promoting effective 'competition between persons engaged in, or in commercial activities connected with, the provision of water and sewerage services'[21] (DEFRA 2005: 1). Significantly, however, when the Ofwat[22] framework was set up, provision for third-party access to sewerage infrastructure was not included, apparently because the industry did not foresee the need for it. Accordingly, the Ofwat framework only applies to water supplies.

In California, legislation known as the Katz Wheeling Law was introduced in 1986. It provides for unused capacity, within artificially constructed water-conveyance facilities, to be made available to others (Slater 2005). However, the legislation does not provide clear guidance on how issues of access pricing should be dealt with and this has acted as a disincentive to third parties pursuing access.[23]

In Australia, there has been only limited experience of third-party access to water. One such example is in the case of Barossa Infrastructure Limited (BIL). That company gained access to transportation capacity in the Mannum–Adelaide pipeline, as well as storage capacity in the Warren Reservoir. The purpose of the access is to provide extra irrigation water from the Murray River for the Barossa Valley. The BIL scheme involved significant upgrading of the SA Water system in order to assist BIL gain a year-round supply of water, and goes beyond a simple third-party access regime (Marsden Jacob 2005: 22).[24]

In Southeast Queensland, there is another arrangement which could perhaps be described as a form of third-party access. It involves Southeast Queensland Water providing raw bulk water which, in turn, is pumped to Brisbane Water's treatment plant and then eventually provided to the customers of Brisbane Water, Logan City Council and Gold Coast Water. Brisbane Water's infrastructure is used by other retailers in this process (Marsden Jacob 2005: 22).

The case of Lakes R Us Pty Ltd represents a thwarted attempt to gain third-party access to water storage and transport services provided by Snowy Hydro Limited and the State Water Corporation.[25]

Given that state-based third-party access regimes for water services can be made compliant with National Competition Policy and National Water Initiative goals it seems possible that other states will join New South Wales in introducing third-party access legislation of their own, rather than rely solely on the generic Commonwealth access regime under the *Trade Practices Act 1974* (Cth).[26] The New South Wales specific state-based water-access regime is a 'significant and unique reform', with 'no similar regime elsewhere in the world' (Freehills 2006). It is helpful, therefore, to examine the development of legal frameworks for third-party access, beginning with the Commonwealth scheme.

Commonwealth legal framework for third-party access in Australia

The move to open competition in the water and wastewater sector came with the 1993 release of the Hilmer Report.[27] This argued for greater competition among government-owned entities, the removal of trade barriers, the elimination of cross-subsidies in the provision and delivery of water and the abolition of monopoly practices.

Following the Report and the Competition Principles Agreement, amendments were made to Commonwealth legislation, including the *Trade Practices Act 1974* (Cth). State-based legislative changes followed more slowly and were often tied to Commonwealth incentive payments to the States, which were made in three tranches from 1997 to 2005 to induce State compliance with National Competition Council (NCC) goals (NCC, Principles for Reform).[28]

The *TPA* was amended to include Part IIIA, which relates to 'Access for Services'. Under that Part, third-party access to nationally significant infrastructure may be sought. This may occur by virtue of a 'declaration' of the service by the designated Minister (*TPA*, Part IIIA Divison 2) or by virtue of service-provider access undertakings or industry access codes approved by the Australian Competition and Consumer Commission (ACCC, *TPA*, Part IIIA Division 6). Whilst the designated Minister will normally be the Commonwealth Minister, it is the responsible State or Territory Minister if the service provider is a State or Territory body, such as a public water utility. A service may not be declared, and an access undertaking or industry access code may not be approved, if there is in place an 'effective access regime' established by the State or Territory.[29]

Where the infrastructure concerned is not nationally significant (and, therefore, beyond the jurisdiction of the Commonwealth), state-based access regimes consistent with sub-clauses 6(3) and 6(4) of the Competition Principles Agreement are to be used to gain access.[30]

The *TPA*'s generic scheme for declaration of a service involves a two-stage process. First, the third-party access seeker applies to the National Competition Council for a recommendation that the designated Minister declares that the infrastructure is 'covered' by Part IIIA. The NCC may not recommend, and the Minister may not declare, that the infrastructure is covered unless satisfied that all six assessment criteria are met.[31] The criteria comprise: the promotion of increased competition in at least one market; the economic infeasibility of anyone else developing another facility for the service; the national significance of the facility; that access would not cause undue human health or safety risk; that access was not already part of an effective access regime; and that access would not be contrary to public interest.

The second stage of the process is the determination of the terms of access to the declared service (TPA, Division 3). An access seeker (it does not have to be the entity that obtained a declaration of the service) will endeavour to negotiate the terms of access with the service provider. If the parties fail to reach agreement, either of them may give notice to the ACCC to arbitrate the dispute. The ACCC may determine the access dispute consistent with statutory restrictions that seek to protect existing user rights in the service, including by provision for compensation by the third-party entrant to an existing user who suffers deprivation of a pre-notification right (*TPA* s 44 W).

One additional point to note here is the *TPA* definition of 'service' that may be the subject of an access declaration.[32] It is defined to mean:

'a service provided by means of a facility and includes:

a.   the use of an infrastructure facility such as a road or railway line;
b.   handling or transporting things such as goods or people;
c.   a communications service or similar service;

but does not include:

d.   the supply of goods; or
e.   the use of intellectual property; or
f.   the use of a production process;

except to the extent that it is an integral but subsidiary part of the service.'

It is suggested that this definition, particularly paragraph (d), means that a Part IIIA declaration may apply to compel access to the sewerage transportation infrastructure of a service provider, but may not apply to compel the service provider to provide 'access to the goods'; that is, the raw sewage resource.[33] Thus, Part IIIA could be used to compel access for a competing sewerage-service provider or a sewer miner to the sewerage infrastructure but it could not be used by a sewer miner to compel access to the sewage resource itself.

So far, there has only been one successful wastewater access application made under this Part IIIA regime, and that is Services Sydney's application. This contrasts significantly with the position of third-party access in the gas and electricity sectors (which can also use the *TPA* generic scheme) where, by 2005, 60 arrangements had been registered with State and Federal agencies (Marsden Jacobs 2005: 86). In part, the difference is accounted for by the resistance to the dismantling of the vertically integrated, monopolistic public water utilities, owned by State and Local Government agencies. As well, there are difficulties associated with transporting water, which makes the market structure for water more complex than gas and electricity.

Services Sydney application: An access regime in practice

Services Sydney Pty Ltd initially began discussions with Sydney Water to become a sewer miner but later decided to become a competitor with Sydney Water as a provider of sewage-collection services in Sydney. Rather than build its own sewerage network, it sought access to Sydney Water's sewerage reticulation network to transport the sewage of its customers (to be) to connections with its own pipeline; a pipeline which would ultimately transport the sewage to its own treatment plant. It proposed to use the treated water for non-potable purposes such as agricultural, industrial and domestic uses as well as for environmental flows. Consequently, in 2004, Services Sydney applied to the NCC for a recommendation that the sewerage networks leading to the North Head, Bondi and Malabar Ocean Outfalls be 'declared' under the *TPA* Part IIIA. The NCC recommended to the then Premier, Mr Carr, that the services be declared, but the Premier did not publish the declaration within the 60-day time period and was deemed to have refused to declare the services (under *TPA* s 44 H (9)). Services Sydney successfully appealed that deemed refusal to the Australian Competition Tribunal (ACT), which declared the service for a 50-year period from December 2005.[34]

However, after declaration a dispute ensued over the access-pricing methodology to be used by Sydney Water and, on 6 November 2006, Services Sydney activated the arbitration mechanism when it notified the ACCC of the dispute. In July 2007, the ACCC released its final determination on the dispute. The ACCC largely found in favour of adopting Sydney Water's 'retail-minus' proposal for calculating the cost of transportation of the sewage by Sydney

Water. It also retained postage-stamp pricing[35] for all customers (ACCC Arbitration Report, July 2007).[36] Having lost the battle to prevent access[37] it was presumably a consolation to Sydney Water that the pricing decision was favourable to them.

Sydney Water's resistance to granting access raises a number of questions, the key of which is, why? Was there good reason to resist the opening up of infrastructure to third parties or was it simply a case of protective self-interest?

There are several possible answers to this question. In its submissions against the NCC draft recommendations, Sydney Water argued that it supported appropriate market structures and competition reforms but believed that access should be considered in the 'context of an overarching market framework for providing water and wastewater services' rather than be driven by specific access proposals (NCC Draft Recommendations 2004: 3). Put another way, Sydney Water wished the issue to be approached holistically, so as to ensure that market structure and access arrangements supported Governmental policy and facilitated decisions relating to consumption and production (NCC Draft Recommendations 2004: 3). Sydney Water recognised its obligation to find ways of using existing potable water supplies more efficiently and of finding alternative sources that represented 'least-cost' outcomes for the community at the current level of service. It believed that those goals did not necessarily sit easily with private investment, which is more likely to be driven by profit-maximisation objectives.

Sydney Water was also concerned that no appropriate access-pricing arrangements had been resolved. Sydney Water acknowledged that, if a declaration were made (as it ultimately was), the then current integrated wastewater tariff would need to be 'unbundled' in order to determine a separate price for the use of the sewer networks. However, it believed that the then current integrated wastewater tariff, as determined by IPART, did not reflect the full cost of retail sewage-collection services. As a result, Sydney Water considered that it would not be economically viable for a third party to enter the dependent market without some form of subsidy.[38]

A further reason for Sydney Water's resistance to third-party access may have lain in a concern that Australia and, in particular, New South Wales were entering unchartered territory by supporting the Services Sydney application. The Services Sydney plan was, and remains, quite novel and consequently means that it is difficult for Australia to call on the benefit of others' experiences. Given that the resource over which the regime is to operate (water) is a very precious, life-sustaining one, the stakes seemed alarmingly high. In fact, to many they still do.

Yet other reasons for Sydney Water to oppose the application so fervently included:

a.   the cost to Sydney Water of making the significant changes to facilitate the obtaining of access by a third party;

b.   concerns about the provision of adequate consumer protections such as providers of last resort in the case of emergencies;

c.   concerns about how third-party access would fit with obligations to maintain environmental standards, particularly as they relate to leaks and spills;

d.   concerns about public safety and performance standards; and

e.   the constraints that third-party access may impose on future reform of the water sector, such as the discouraging of new entrants into franchise markets because of concerns about the number and nature of contestants in the retail market (NCC Draft Recommendations 2004: 4).

Services Sydney's application under this regime has proved very expensive and time-consuming. Three years after lodging its original application, it still has not actually gained access. Only in 2007 has it learnt of the methodology that will be used to calculate an access price.[39] However, it is on its way to establishing a competing sewerage service and gaining access to that new resource — sewage.

The *Water Industry Competition Act* 2006 (NSW)

While resistance to making a declaration under the *TPA* continued, the New South Wales parliament worked on the development of a state, industry-based access scheme, which is now housed within the *Water Industry Competition Act 2006 (WICA)*.[40] This Act is examined as a case study because some other states are also considering the introduction of legislation setting up state-based access regimes.[41]

WICA is designed to reflect the State government's *2006 Metropolitan Water Plan* [42] and is based on recommendations in the 2005 IPART review of water and wastewater service provision (IPART *Water and Wastewater Report*, October 2005). The IPART review included recommendations for:

1.   the State government to develop a state-based access regime for services traditionally provided by significant water and wastewater infrastructure;

2.   a review of the legal framework to ensure appropriate obligations were placed on both incumbents and new entrants, as well as ensuring that barriers to competition and private-sector involvement were addressed;

3.   the development of a streamlined regulatory framework for sewer mining which included establishing a formal dispute-resolution process and;

4.   continued IPART price regulation of services to small customers, as well as the continuation of price regulation for large customers, regulation which would be reviewed if an access framework were established.[43]

The Act was designed to encourage competition in the water industry and to promote innovative solutions to the water supply-and-demand balance, particularly in so far as the development of infrastructure for the production and distribution of recycled water is concerned.[44] WICA is, in many ways, only a skeletal framework which needs to be fleshed out by operational and enforcement mechanisms. The Act contains provisions for:

a.   a Licensing Regime for private-sector participation in the water industry (in Part 2);

b.   a third-party access regime to facilitate the negotiation of access to significant water-industry infrastructure (in Part 3);[45] and

c.   binding arbitration of sewer-mining disputes (in Part 4) (DWE, *WICA Regulations Consultation Paper* 2007: 8).

The detail for the administration of these Parts is provided by the Regulations made under the Act. Section 101 provides for the Governor to make Regulations in respect of a range of issues.

Licensing regime

The legislation includes a licensing regime, rather than alternatives such as rules-based regulation, certification or industry self-regulation because the licensing regime was thought to offer the best adaptive capacity, allowing flexible responses to future situations that competition, public expectation and scientific knowledge might present. Accordingly, private entities that seek to provide drinking, recycled and other grades of water, *as well as those seeking to provide sewerage services,* will need to be licensed. The licensing requirement operates by virtue of a prohibition-and-exemptions approach. The Act makes it an offence to construct, maintain or operate any water-industry infrastructure or supply water or provide a sewerage service by means of any water-industry infrastructure without a licence. There are monetary penalties for non-compliance (*WICA* s 5).

The Act also establishes a set of principles that are to guide the Minister when he or she is deciding whether to grant a licence. Those principles relate to issues such as the protection of public health, the environment, public safety and consumers; the encouragement of competition in the supply of water and the provision of wastewater services; the ensuring of sustainability of water resources; and the promotion of production as well as the use of recycled water (*WICA* s 7). Unfortunately, the constraints of this chapter do not permit further exploration of these issues. (See DWE *WICA* Regulations Consultation Paper 2007.)

Access regime under WICA

Central to this chapter is the third-party access regime under Part III of *WICA* (MWDCO, *Creating a Dynamic and Competitive Water Industry* 2006: 18). The access regime is designed to operate in concert with the licensing regime, in that water and wastewater-services providers gaining access under the access regime will also need to be licensed in order to operate. The aim of this industry-specific access regime, like the Commonwealth's generic one under the *TPA*, is to facilitate third parties gaining access to the incumbent's 'infrastructure service', thus allowing third parties to become new service providers in the upstream or downstream markets without having to incur the expense of duplicating the infrastructure.[46] Third parties may be interested in seeking access so as to supply drinking water, recycled water or wastewater services, for example.

The regime specifically facilitates the negotiation or arbitration of arrangements for third-party access to the storage and transportation facilities of the incumbent's water-supply and sewerage networks. The definition of 'infrastructure service' (*WICA* s 4 and dictionary) as the 'storage, conveyance or reticulation of water or sewage by means of water industry infrastructure' but not including 'the supply of goods (including the supply of water or sewage)' shows that the access regime itself does not create rights for the access seeker to obtain the raw resource of sewage. However, it is anticipated that the execution of a Sewer Network Access Agreement between the sewerage network owner and the access seeker will not only create rights to extract a designated volume or proportion of sewage but will cause that agreed volume or proportion to be extracted.[47] Further, *WICA* does not create rights to use the incumbent's treatment plant.

Initially, the third-party access regime is only available to access seekers in the Sydney and Hunter regions but that could be extended (WICA s 22 and Schedule 1 of Act). An infrastructure service would become the subject of the access regime if:

1. the Minister makes a 'coverage declaration' in respect of it;[48] or
2. a service provider gives an 'access undertaking' in respect of it (*WICA* s 38.)

Where a service provider has given an 'access undertaking' for the infrastructure service, IPART must still approve the access on the basis of statutory criteria in cases where the undertaking is lodged with it (*WICA* s 38(a)–(d)). Once approved, the access provider is required to negotiate in good faith to accommodate the access seeker's requirements. If commercial negotiations break down and agreement is not reached, the matter may be referred to IPART for arbitration.[49]

The access seeker will need to request that the Minister declare the service covered by the access regime where: (a) the service provider has not agreed to access by suitable private arrangements (that is, without lodging an access undertaking with IPART) and the service provider has not given an access undertaking which has been lodged with IPART; or (b) the service has not been previously been declared. IPART will advise the Minister on whether he or she should declare the service.

A declaration will only be made if the service meets the 'declaration criteria' in *WICA* s 23. They reflect those in the TPA's generic model and require that the infrastructure be of State significance; it would not be economically feasible to duplicate the infrastructure; access is needed to promote a material competition in an upstream or downstream market; the safe use of infrastructure can be ensured at an economically feasible cost and; access would not be contrary to the public interest. A service will not be the subject of a coverage declaration if it is subject to a binding non-coverage declaration (*WICA* s 25(5) (a) and Division 4), subject to a voluntary access undertaking (*WICA* s 25(5) (a) and Division 5) or IPART has determined, with the consent of the Minister, that it is a frivolous or vexatious application (*WICA* s 25(5) (b)).

If a declaration is made, the requirement to negotiate the terms of access using all reasonable endeavours will be triggered. If the commercial negotiations as to terms break down, the matter may again be referred to IPART for arbitration (*WICA* s 40).

In practice, the issue of access pricing will play an important role in the effectiveness of the access regime. This is discussed later in this chapter.

Sewer Mining Disputes under WICA

Part 4 (ss 45 & 46) of *WICA* creates a mechanism to resolve disputes about sewer-mining access, both as to the terms of a proposed agreement and as to the application of an established agreement. It provides for disputes between the sewerage-service provider and the sewer miner to be arbitrated by IPART or a person nominated by IPART, but only if the service provider has lodged a notice with IPART setting out the provider's policy on sewer mining. The arbitrator must give effect to the service provider's policy and, subject to that policy, any other matters prescribed by regulation. *WICA* does not require or authorise IPART to review or approve a sewer-mining policy.

The statutory policy would seem to be that sewer mining can only occur with the in-principle agreement of the sewerage-service provider that it is willing to consent to the extraction of raw sewage. The service provider may decide to manage sewer-mining access according to its own policy and entirely outside the *WICA* regime, just as it could before the enactment of that Act. *WICA* does not provide a mechanism to compel the service provider to grant access to the sewage resource itself, as opposed to the sewerage infrastructure. Sydney Water retains control over the amount of wastewater that may be extracted from its sewerage network so that it can ensure the effective operation of the network. That said, Sydney Water's policy is to refer sewer-mining disputes to IPART for arbitration (Sydney Water, *How to Establish a Sewer Mining Operation* 2006: 8).

Yet, it would also seem equally possible for a potential sewer miner, whose activities are thwarted by the infrastructure owner, to achieve its end by going into competition with the infrastructure owner as a competitor in the provision of wastewater services. This method requires that the service is the subject of a coverage declaration and that a price determination has been made for the use of the infrastructure. However, by using this route the third party gains direct access to the sewage resource from the sewerage customers and is able to by-pass the infrastructure owner's resistance to its sewer mining.

WICA's relationship with other legislation

Legislation and codes such as the *Local Government Act* 1993, the *Public Health Act* 1991, the *Protection of the Environment Operations Act* 1997, the *Sydney Water Act* 1994, the *Environmental Planning and Assessment Act* 1979, the Australian Drinking Water Guidelines and the NSW Plumbing Code of Practice 2006 currently provide public health, public safety and environmental protections for the water industry. All service providers, whether they are continuing public utilities or new third-party entities gaining infrastructure access, will be bound to comply with these requirements. Hence, the access regime under *WICA* will operate subject to these prior protections. Technically, there is no question that health, safety and environmental protection standards will be by-passed by the introduction of private-sector, competitive involvement. Indeed, the Regulations Consultation Paper suggests that 'the various legislative and regulatory drivers can therefore be viewed as supplementing the commercial drivers of suppliers to ensure that they deliver services at the level, quality and reliability customers need, at the lowest long-term cost (while also meeting health, safety and environmental obligations)' (DWE, *WICA* Regulations Consultation Paper 2007: 11).

However, critics of water-industry competition question whether this objective is realistically achievable or whether the drive for profit necessarily compromises the ability to deliver services at the lowest long-term cost, while meeting health and environmental standards. They cite Sydney's Cryptosporidium and Giardia scare (which the McLellan report found was probably, in part, linked to problems at the *privatised* Prospect treatment plant)[50] and Adelaide's 'big pong'[51] (which also involved a private entity in the form of United Water) to highlight the tensions of private-sector involvement in the industry. Critics also question whether the wider legislative frameworks beyond

WICA are sufficiently stringent to deal with new problems that may arise as a result of third parties gaining access to public-utility infrastructure, for example. In that regard, it has been suggested that the *Public Health Act* 1991 be amended to introduce new offences prohibiting the supply of water which is a risk to public health and the supply of non-drinking water in circumstances where it could reasonably be mistaken for drinking water (MWBCO 2006, *Creating a Dynamic and Competitive Water Industry*).

The role of Regulations under WICA

The Regulations give the Act further shape, form and substance and the capacity to supplement the public health, consumer and environmental protections of the other legislation. *WICA* authorises the Governor to make Regulations pertaining to water quality and public health; construction and maintenance of water-industry infrastructure; consumer protection; licensing administration and licence conditions, for example (*WICA* s 1001 and Schedule 2).[52] The potential effect of regulations on these issues will not be explored here.

However, it is opportune to note a draft Regulation applying specifically to the access regime. The NSW Cabinet Office released its draft Water Industry Competition (Access to Infrastructure) Regulation in mid September 2007.[53] The draft nominates the persons whose submissions are to be invited in relation to coverage applications for an infrastructure service and these persons include the incumbent service provider along with various relevant Ministers. The Regulation also sets out the process for negotiating disputes concerning access determinations. It places an onus on the service provider to use all reasonable endeavours to accommodate the access seeker's requirements but interestingly does not require those requirements to be reasonable themselves (NSWCO, *Access to Infrastructure Services Regulation* 2007: Div 2, 8 (3)). Further, it sets out a timeline for dispute resolution and outlines the basis on which disputes may be determined. The Regulation does not include a definition of 'capacity' or 'spare capacity', presumably recognising these as site-specific.[54]

Principles of Regulation

As part of the process of developing the *WICA* Regulations, the New South Wales government endorsed some best-practice principles of regulation (DWE *WICA Regulation Consultation Paper* 2007: 11). One of those principles is that of periodic review and reform. Given the changing landscape of regulation, this is welcome.

Regulatory models are diverse and may embrace approaches including 'centred' regulation, a 'hollowing out of the state'[55] or 'smart' regulation amongst others.[56] Although a discussion of regulatory theory and practice is well beyond the scope of this chapter, it is perhaps useful to observe, as does Julia Black, that regulation has developed into a multi-layered phenomenon which may be transnational, supranational, national or sub-national, for example. Regulation is 'thus not simply an activity carried out by the state using laws backed by sanctions; it is a broader enterprise consisting of a sustained and focused attempt by state or non-state actors to alter the behaviour of others with the intention of producing a broadly identified outcome or outcomes'.[57] This will need to be borne in mind when assessing *WICA*'s effectiveness as a regulatory regime, particularly in relation to how well it deals with issues of proportionality, accountability, consistency and transparency, as well as its ability to target areas of risk.

Key issues of interest

Various key issues emerge in relation to the operation of the access regimes under the *TPA* and *WICA*; namely, coverage declarations, access and resource pricing, building trust, step-in rights for operators of last resort, property law issues about the ownership of sewage, and customer and consumer protection. They are treated discretely below, with a focus on the implications for the *WICA* regime.

Coverage declarations

In many ways, the *WICA* access regime reflects the generic access regime available under the *TPA*. For example, the criteria for declaring an infrastructure service open to the access regime are quite similar. Importantly, both include requirements that access or increased access would 'not be contrary to the public interest'.[58]

However, retention of the public-interest test in its present form has prompted criticism for not placing a positive onus on the access seeker. To explain, the *TPA* does not define public interest and the NCC and the ACT[59] have dealt with the issue on a case-by-case basis, relying tentatively on clause 1.3 of the Competition Principles Agreement (PIAC *Submission to Creating a Dynamic and Competitive Water Industry* 2006: 11). That Agreement supports considerations relating to ecologically sustainable development, social welfare and equity considerations, occupational health and safety issues, industrial relations issues, economic and regional development, consumer interests, competitiveness of business and the efficient allocation of resources being taken into account. Accordingly, the Public Interest Advocacy Centre (PIAC) distilled from those considerations issues such as 'the loss of equity in pricing, the impact on prices for the incumbent, risk of consumers being excluded from price benefits and the magnitude of public costs' as all being relevant and coming under the head of 'public interest' (PIAC submission to *Creating a Dynamic and Competitive Water Industry* 2006: 11). PIAC went on to argue that the *WICA* access regime should place a greater burden on the access seeker to demonstrate that these concerns have been adequately addressed. It also favoured demonstration of a positive-outcomes approach by access seekers rather than the mere demonstration that outcomes will not negatively affect the public interest. PIAC claimed that its proposal was consistent with the Intergovernmental Agreement on a National Water Initiative (IGA), which was agreed in June 2004 by the Council of Australian Governments (*PIAC Submission to Creating a Dynamic and Competitive Water Industry* 2006: 13). It favoured the onus in the public-interest test from negative to positive.

While at one level it would seem insufficient for access seekers to argue that their business models promote competition and, by extension, that environmental benefits will flow and that those benefits will not be against the public interest, change to the public-interest test along the lines of PIAC's recommendation may also pose problems. The proposed switch of onus from negative to positive could potentially place an enormous burden on the applicant unless the public-interest test were focused on achieving better outcomes on specified issues, such as pricing and environmental protection.

In regard to retail pricing, the applicant already has to show how access will promote competition. One of the difficulties in showing that competition will reduce prices is that the public-utility models have artificially suppressed prices by applying a pricing formula which has not been based on full cost recovery. The Australian Competition Tribunal (ACT) in the Services Sydney decision acknowledged the difficulties caused by artificial price suppression (*Application by Services Sydney Pty Ltd* [2005] at [201]).

As to environmental outcomes, any likely adverse environmental impact from an associated works proposal would be subject to an environmental impact assessment. Hence, protective mechanisms are already in place. Yet, perhaps PIAC would consider that mechanisms of this nature still do not go far enough because they do not require demonstration of a positive outcome. They simply protect against adverse ones.

In the specific case of Services Sydney, it is arguable that the company could have easily satisfied a positive-outcomes test because it proposed treating the relevant sewage to secondary and tertiary levels (*Application by Services Sydney Pty Ltd* [2005] at [70]–[73]), whereas Sydney Water had only been treating sewage to primary levels and then discharging it to ocean outfalls. Further, Services Sydney also intended charging customers much the same price as Sydney Water charged, but Services Sydney planned to attract customers with its greener credentials. The problem in such a case is perhaps not one whereby the applicant finds meeting the positive-outcomes test too burdensome but, rather, that there would be little incentive for the public utility to improve its own standards when to do so would have cost implications for it.

Access and resource pricing

Questions have emerged about the pricing of access to the infrastructure service and of access to the sewage resource itself. The pricing of both may prove important to setting the conditions for successful entry of third parties into the wastewater and water-recycling industries. The power to determine the price for access to infrastructure rests with the ACCC under the *TPA* and with IPART under *WICA* where the incumbent service provider and the access seeker cannot agree (see previous discussion). The power to determine the price of recycled-water services (that is, the supply of recycled water) rests only with IPART under the *IPART Act* 1992 (NSW) s 11 and Part 3, Division 5, but only in respect of prices to be charged by a 'government monopoly service'.

What of the power to determine the price of the sewage charged to the sewer miner? On our reading of the *IPART Act*, IPART also has the power to determine the price of the sewage resource charged by a government monopoly service to a sewer miner. However, an alternative opinion is that the *IPART Act* only authorises IPART to determine the price for the interconnection service for sewer mining, which does not include pricing the sewage resource itself. However, under *WICA* ss 45 and 46, IPART does have the power to determine the price of the sewage to be drawn by a sewer miner where the infrastructure 'service provider' has agreed to IPART having the jurisdiction to arbitrate such disputes consistent with the terms of the service provider's policy on sewer mining. That policy may contain propositions about pricing the sewage. Here, however, the definition of 'server provider' means any person who has, or is to have, control of the infrastructure; so this will include private providers. Thus, the service providers, public and private, can set the parameters for IPART's determination of the price of the sewage as a resource.

The price of access to the incumbent's infrastructure is a very important factor in the success or otherwise of any third-party access regime. Should the price of access be set too high, competition is unlikely to ensue because access seekers will presumably opt not to take up the access opportunities that they have been granted. Alternatively, if access seekers do take up the access that they have been granted and enter the market but find that the access price has been set at a level too high for them to operate a successful business, the customers of the incumbent and the new service provider may ultimately pay for inefficiencies in price setting. A possible outcome is that one group of customers may end up subsidising another (PIAC Submission 2006 *Creating a Dynamic and Competitive Water Industry*: 11).

The legislation provides various principles to be applied in determining the price of access, but there are still choices about pricing methodologies that are available (*TPA* s 44X; *WICA* s 41). Some were discussed in the ACCC's *TPA* determination of the access-pricing dispute between Services Sydney and Sydney

Water (ACCC *Arbitration Report* 2007). The ACCC noted that the most appropriate methodology depends on a range of factors, including the infrastructure facilities to which access is sought and the particular characteristics of the upstream and downstream markets (ACCC *Arbitration Report* 2007: 1).

In that dispute, Services Sydney favoured a 'bottom-up' building-block methodology under which the price is calculated by building up the various blocks of costs associated with providing the service. Sydney Water, on the other hand, favoured a 'retail-minus' methodology, which calculates price by first determining Sydney Water's regulated retail price and subtracting from it the cost of contestable services (for example, sewage treatment and sewage disposal and recycling) associated with the supply of the product or service in the downstream market. To that figure any facilitation costs (those associated with the provision of the service by Sydney Water to Services Sydney) are then added.

Ultimately, the ACCC favoured Sydney Water's 'retail-minus' approach but, instead of subtracting 'avoided' costs, it subtracted 'avoidable' costs or costs that could in the long run be avoided. The result is that access prices are lower than if only actually avoided costs were subtracted. The types of concerns addressed by the ACCC under the *TPA* regime will also need to be addressed in relation to access that is made available under *WICA* and other state-based regimes (Economic Regulation Authority, Western Australia, 2007: 79 ff).

Pursuant to *TPA* s 44X(2), the ACCC also took into account some other matters such as the complexity that would be involved in practically implementing each party's proposed access-pricing methodologies. As part of its decision, the ACCC retained postage-stamp pricing for all customers, thus favourably addressing many equity concerns raised by consumer-interest groups. Although the provision of sewerage services to different parts of Sydney costs vastly different amounts, to a considerable extent consumers have already paid for that cost differential through the developer charges for different areas, which are meant to reflect the cost differential. To apply this cost differential to the periodic charges for sewerage services would effectively require the consumers in high-cost areas to pay twice.[60] Consequently, *WICA* specifically requires that the pricing principles for access to infrastructure services be implemented in a manner that is consistent with any pricing determinations for the supply of water and the provision of sewerage services, including the maintenance of 'postage-stamp pricing' for the provision of those services.[61]

However, a real concern for legislators is that, having developed extensive regimes for access, the so-called opportunities might not be taken up by third parties because the price of access does not leave the opportunity for sufficient profit. Anecdotally, some large water companies have expressed concerns about whether it is worth their while entering the market. They have suggested that, due to cost constraints, third-party access may well end up only being viable for greenfield residential developments or large industrial sites. In Sydney, this would limit third parties to about a 20 per cent market share, given that it has been estimated that about 80 per cent of Sydney Water's existing customers are not in greenfield developments and are small, largely residential customers. If that is so, perhaps the *TPA* and *WICA* third-party access regimes might not provide the levels of competition envisaged by legislators. Further, with retail prices for services being set by state regulators (for example, by IPART in NSW) and the infrastructure access prices being set by the ACCC in cases coming under the *TPA*, some third parties may see high transaction costs as a disincentive to invest. Practitioners report that there does not appear to be a large number of potential, third-party access seekers lining up to enter the market.[62] Yet this position could change. Indeed, a declaration made under the *TPA* could even be revoked under s 44J of that Act if an alternative, effective state-access regime becomes available. One would expect that the *WICA* regime should provide the basis for the *TPA* declaration of Sydney Water's sewerage network to be revoked. That expectation should be measured against the sceptical assessment of *WICA* by one NSW member of Parliament: 'The Competition Bill may well prove to be only window-dressing in terms of actual competition.'[63]

It will be interesting to see whether in the hard, cold, light of a commercial day Services Sydney ultimately decides that it is worthwhile pursuing the competitive opportunity available to it, given the price it will have to pay for access to the infrastructure service of sewage transportation.[64] Much may depend on (a) the estimates of the value of the ultimate product — recycled water — and (b) the related cost of acquiring the raw resource of sewage in the first place. It would seem that Services Sydney may not have to pay for the resource at all. As Sydney Water would not be the supplier of raw sewage to Services Sydney, it could not charge for supply of the resource as distinct from the provision of the service of sewage transportation. Instead, Services Sydney will take delivery of the sewage from its own customers. If Services Sydney customers 'give' rather than sell their (purported) property rights[65] in 'their' sewage to the company, then the company will receive a windfall benefit of 'free' sewage.

Accordingly, in the recycled-water market, Services Sydney may be at a competitive advantage over sewer miners who take delivery of raw sewage from Sydney Water. Why? Because IPART chose not to regulate sewer-mining prices in its 2006 pricing determination (IPART, *Pricing Arrangements for Recycled Water and Sewer Mining*, September 2006: 66). Unless IPART regulates the price of sewage for sewer miners, Sydney Water will be able to charge them what it likes if it abandons its current policy of zero pricing of sewage. Even more significant, potentially, is that there is no regulatory regime applicable to the choices that private sewerage-service providers make with respect to sewer miners, either as to whom they choose to sell the sewage or as to the price that they charge. Similarly, private sewerage-service providers who decide to recycle the water will be subject to no direct pricing regulation for their product, other than the market competition from the regulated public-sector providers. If public-policy concerns develop about how private providers are allocating the sewage resource, either as sewage or as recycled water, then consideration may need to be given to whether sewage may need to be made the subject of a resource regulation regime, not just a competition and service regulation regime. These issues are ripe for further investigation.

Building customer trust

In the New South Wales electricity and gas sectors, enticing customers to move to new third-party competitors has been partly dependent on making customers feel confident about entering into negotiated contracts in the first place.[66] It would seem that establishing the requisite levels of trust and comfort in those sectors may have posed some difficulties because the 'churn' rates in relation to residential electricity supplies have remained relatively low. ('Churn' is the rate of switch between competing suppliers and is the basic measure of retail competition.) The rate of churn for residential electricity customers in New South Wales was, according to the Public Interest Advocacy Centre, approximately 15 per cent and that was after four years of retail competition.[67] Hence 85 per cent of retail customers remained with the incumbent supplier and purchased electricity at the regulated price. In the gas industry the position was similar.[68] After approximately five-and-a-half years the churn rate has increased but the competitor's market share remains relatively small. About 70 per cent of electricity customers have opted to remain with the incumbent supplier according to IPART's 2007 figures (IPART, *Overview of Final Report, Retail Prices in Electricity in NSW* June 2007: 1). It is possible that this position would not be entirely dissimilar in the water sector.

It is acknowledged that in some other jurisdictions, such as South Australia and Victoria, the churn rates in the electricity sector have been higher than in New South Wales (PIAC Submission to *Introducing a Dynamic and Competitive Water Industry* 2006: 5). However, IPART also acknowledged that New South Wales would be unlikely to approach these high churn rates because the rate of switch in Victoria and South Australia was associated with a peculiar set of circumstances.[69] IPART also noted that the average switching rate for (electricity) customers in the European Union was around 10 per cent and that in New Zealand, which had the longest history of contestability, the churn rate was also around 10 per cent (IPART, *Promoting Retail Competition* Final Report: 104). This suggests that churn rates are not generally high. Further, although IPART has proceeded on the basis that its determination is likely to lead to increased competition, it admits to still being uncertain about whether that will translate into increased churn rates. IPART has expressed the view that increased rivalry between firms may have positive spin-offs for existing customers, in the form of discounts or innovative services, for example, but not impact greatly on churn rates (IPART, *Promoting Retail Competition* Final Report: 105). Whether this is ultimately the case is not yet known but recent debates surrounding the potential privatisation of the electricity sector in New South Wales suggests that IPART's confidence in the benefits of competition is, at least, contestable.[70] It may be helpful to bear in mind the experience of the electricity sector when assessing competition in the (waste)water sector, albeit that there are some notable differences between the two.

Perhaps one reason for the relatively low churn rate in the New South Wales electricity sector is customer resistance to being contacted (particularly by telephone) by competitive operators offering 'better deals'. The implementation of the national 'Do Not Call Register'[71] is just one measure of how far customers will go to avoid being contacted by retailers seeking to sell 'their product'. Hence, it may be more problematic than first thought for new third-party competitors in the wastewater sector to attract customers. (Dovers discussed the significance of human behaviour and conduct in Chapter 5 when he observed that although 'we talk of "water management" ... it is really about managing people'.) Ease of access to the customer base, plain-English information, positive financial and service benefits, as well as confidence and trust in the retailer itself, will be important factors to address if the introduction of alternative wastewater-service provision via third-party access to infrastructure is to succeed. The introduction of 'retail headroom' in pricing may achieve a greater take-up rate of alternative service provision but the *quid pro quo* could be financial hardship for many customers.[72]

Step-in rights for operators of last resort

How will the regulatory framework deal with the delivery of water or sewerage services if that delivery is prevented because of physical problems with transportation or infrastructure, insufficient water, the financial distress of the operator or the financial distress of the retailer? One mooted solution is the creation of 'step-in' rights for replacement operators, often known as 'operators of last resort'. This approach is common in the electricity sector, for example, where electricity can be fairly readily transferred between grids. However, such transferability is not so simple in the water sector. Indeed, Bakker described water as an unco-operative commodity (Bakker 2003). In practice, therefore, one could imagine that those nominated as 'operators of last resort' would not necessarily be poised, ready to step in, should they be needed. The task of supplying the water or wastewater service may well be limited by physical and technical constraints. Operators of last resort would presumably also have their own businesses to run and may be diverted by activities relating to those obligations at the time they are most needed to step in.

A related concern is whether the new private-sector operators (some of whom may have third-party access to infrastructure) will bring with them the technical and management skills to operate their businesses successfully. This is of particular concern where they are entering new fields. Inexperience bears its own inherent set of problems. To explain, one wonders what would be the case if a private operator faced technical problems which it could not solve and so needed assistance from the public utility, who was the operator of last resort, but the expertise traditionally housed in the public utility operator of last resort, had been lost or depleted when its staff were retrenched as a result of a private competitor entering the industry?[73] Anecdotally, this kind of scenario has been used to explain the position in Adelaide at the time of the 'big pong'.[74]

There is also the question of risk-auditing in relation to the payment of bonds under the licensing regime to offset 'step-in' costs in the case of an emergency. A scheme for bond payments would require assessments to be made about operators and that, in turn, would raise questions about where the risk should be parked. Should it be placed with the community at large, or should it be borne by the retailer, the end-user or others? Under socio-liberalism, the entrepreneur bore the risk but under neo-liberalism we have seen a shift to democratise risk and have it borne by the whole of society, while profit has remained largely privatised. How best to deal with these issues is something that legislators will need to review constantly in relation to the water sector.

Property law issues

The issues of third-party access, wastewater service provision, sewer mining and recycling also raise some fundamental property-law issues that deserve further consideration. There is already, in other contexts, established law dealing with the ownership of, and access to, infrastructure, such as the pipes which lie on private land.[75] That law is based on statute (for example, *Sydney Water Act 1994* s 37), the doctrine of fixtures, easements and licences. *WICA* also deals with the question of ownership of a licensed network operator's water-industry infrastructure, deeming it to belong to the operator (*WICA* s 64). The more interesting question is the ownership of the sewage itself. This is particularly pertinent in the context of sewage being cast as a valuable asset. It excites interest in the related question of which party should be reimbursed, if any, for the sewage used to make profit through on-sold recycled water or treated sludge.

While water as a flowing natural resource is not the subject of property,[76] water in other circumstances may be. For example, trade in bottled spring water, treated tap water or 'manufactured' effervescent water all demonstrate the ability of water to be commodified and become the subject of property. Hence, an argument exists that a householder who pays for water to be supplied by a retailer, such as Sydney Water, purchases the water that is supplied. The water becomes his or her property. Subject to some restriction (for example, Sydney Water's restrictions on hosing), the water may be used and enjoyed and transferred to another, while third parties may also be excluded from it. The householder may use the water to take a shower, boil some potatoes, wash clothes or mix with some lemons to make and sell as lemonade. Should the water have been captured by rainwater tanks on the householder's property instead of purchased from a service provider, the conclusion that the water has become the householder's property applies *a fortiori*. The householder, in capturing the water, has brought it into his or her possession and brought it under his or her control.

If the water is the property of the householder who purchased (or captured) it, this, in turn, raises the question of whether the water becomes any less his or her property once it has been despoiled by human and other waste and flushed into the sewerage system. To explain, the food scraps that ultimately flow into the sewerage system when the sink plug is released are certainly the property of the householder before they enter the sewerage system. Human bodily waste is also presumably the property of the householder before its entry into the sewerage system, although cases such as *Moore v The Regents of the University of California* raise important and interesting issues on the question of characterising body parts as property.[77] But do these proprietary rights terminate once the water is discarded?

On one analysis, the wastewater or sewage remains the property of the householder, who has merely contracted with the wastewater-service provider to perform only a service in relation to it. But that argument has its weaknesses, not the least of which is establishing that the householder/owner intended to continue exercising dominion over the wastewater once it entered the sewerage network. On another reading, the individual householder's title may be said to be transferred to the wastewater-service supplier, directly or indirectly, by way of the wastewater-service contract[78] or any relevant legislation which governs the entity supplying the wastewater service (for example, *Sydney Water Act 1994*). Which view is better will depend on the terms and conditions of the contract. It is possible that the wastewater-service provider is only that, a 'service' supplier. Its role is to supply a service rather than acquire property rights. Hence, although it contracts to remove wastewater, on this line of reasoning it is unlikely to acquire property rights by way of simply performing the service.

Yet there is a weakness in the proposition that the wastewater-service supplier only provides a service in taking away the wastewater. It may mean that the water retailer may also only provide a service when it supplies clean water in the first place. If that is so, the householder would have no property in the water which has become the subject of a dispute. However, this may not be fatal. The householder's proprietary rights in the water may not be dependent on a transfer of property rights from the supplier to him or her. Perhaps the property rights in question are created by the manner in which the householder uses the water and how effectively he or she demonstrates possession and exercises dominion over it.[79] (This is demonstrated even more clearly in relation to the person who collects water in the rainwater tank.)

Yet another, and perhaps more fruitful, way of looking at this question is through the legal doctrine of abandonment. This argument is based on an understanding that, whilst the householder may have held property rights in (a) the water supplied to him or her and (b) the waste matter with which he or she polluted the clean water supplied to him or her, once the sink plug is pulled or the toilet flushed, the householder has abandoned any proprietary rights that he or she may formerly have held.

The doctrine of abandonment involves ownership of a chattel being divested in circumstances where it can be shown that: (a) the original owner intended to renounce his or her title and (b) the chattel lawfully fell into the possession of another.[80] Perhaps it could be argued that when the householder flushes the toilet, he or she is divesting him or herself of any title (by way of the doctrine of abandonment) to the wastewater which he or she might have had in it, rather like the householder who takes his goods to the local rubbish tip and leaves them there.[81] Once the householder dumps his or her goods, the tip operator acquires a possessory title to them by means of the strict control the operator exercises over the tip site (K. and S. Gray 2004: 55). It is possible to conceive of the householder flushing the toilet in a similar manner. Accordingly, once the toilet is flushed, the householder relinquishes his or her title and the wastewater becomes the property of the sewerage-network operator, who can resist the claim of a householder if he or she later tries to claim title over his or her sewage.

The recent decision of *R (Thames Water Utilities) v South East London Division, Bromley Magistrates' Court* dealt with the issue of whether wastewater which had accidentally escaped from a sewerage system constituted waste within Directive 75/442 EEC. In that case, it was alleged that the Thames Water Utilities deposited untreated sewage constituting 'controlled waste' on land in the county of Kent, as well as into controlled waters in that county.[82] The Court found that, although the escape of sewage was accidental, it did not preclude it being 'discarded' (by the Utility) and, in turn, from being characterised as 'waste'. However, the case did not discuss whether waste was inside or outside the property paradigm and nor did the Court decide the issue by reference to the language of property law. Instead, it referred to the 'producer' of waste and the 'holder' of waste. This would tend towards waste being seen, at least in the context of an accidental discharge and the relevant European Community Law, as outside property law.[83] Perhaps this conclusion is supported by the fact that authorities such as Sydney Water do not currently charge for the sewage that a sewer miner removes from the network. This suggests that, to date, sewage has been regarded as worthless, something that could not be given away and, accordingly, there has been little incentive to conceive of it as property or to bring it within the parameters of a trading regime valid at law. However, that may well change if the resource of sewage becomes the subject of pricing. If a price is attributed to raw sewage and commodification follows, a Benthamite analysis would suggest that property will be born. 'Property and laws were born together and die together. Before laws there was no property; take away laws and property ceases.' (Bentham in Macpherson 1978: 52).

If property can exist in 'thin air', as it does in strata schemes, for example, there would seem to be little reason preventing wastewater from being construed as property.[84] Yet, the next question is, whose property is it? Is it the third-party accessor's, the sewer miner's, the incumbent wastewater-service provider's or the householder's, for example?

Further, if individuals find that sewage is valuable, this may result in them wishing to retain it as 'their property' rather than 'giving' it to wastewater-service providers, third-party access seekers or sewer miners. One possibility is that the desire to retain sewage could result in a trend towards individual households or neighbourhoods seeking to set up their own infrastructure to treat sewage in order to enjoy the benefit of recycled water or treated sludge themselves. Governments and legislators may need to consider the implications of this.

We also could find that sewer miners, third-party access seekers, infrastructure owners and individual households may all end up competing with each other for sewage. Sewer miners, such as local councils, may want sewage to recycle, with a view to using the recycled water in public spaces, such as parks and golf courses. Third-party access seekers who wish to pursue recycling businesses may wish to acquire sewage so that they can on-sell recycled water and treated sludge. Incumbent wastewater-service providers may wish to retain the sewage they collect and either enter into or expand their recycling activities while individual households and local communities may wish to retain sewage so that they can recycle it at a micro level. If so, all these parties could conceivably be in competition for the resource. However, this scenario is not likely to occur unless it seems a profit can be made out of entry into the market. At the moment, concerns about access pricing for third parties seem to be causing would-be investors to display timidity and tentativeness about entry into the market.

Of course, another scenario completely is that legislators may decide that there are strong policy reasons not to construe sewage as property 'owned' privately but, rather, to characterise it as a right akin to either *res nullius* (property belonging to no-one) or as *res commune* (property belonging to everyone). On one view, the present Australian passion for privatisation and the commodification evident in the water sector would seem to make this unlikely. However, on another view, if the assertion of proprietary rights by water utilities (public or private) were driving up the price of sewage as a resource and distorting the allocation of recycled water in times of water scarcity, then governments and parliaments may see the public-policy benefit in legislating to assert public ownership of sewage and regulate the allocation of this wonderful new resource!

Negotiating inexperience and customer protection

Another issue is whether the net has been cast widely enough to provide protections for smaller customers, on the basis of their negotiating inexperience. Smaller customers are probably more likely to lack information, be commercially inexperienced and not be equipped with a strong bank of negotiating skills. As a consequence they may find it more difficult to negotiate better alternative-supply arrangements (for services provided by third-party accessors). That would suggest that smaller customers should receive more protections than large consumers; however, public water utilities in the Sydney and Hunter regions currently must offer the same protections to *all* customers alike. One possible problem in creating two classes of customers, small and large, is where to draw the line between them and how to address the concern that some small customers might be well equipped to negotiate and other larger companies might be either inadequate or inexperienced as negotiators. Given that, it may be better for new, private suppliers to be forced to provide the same protections for all their customers. Such a method would ensure that more vulnerable customers, irrespective of size, received protection. It would also streamline the obligations of public and private service providers.

Leases and consumer protection

A further, very practical, problem arises in relation to *WICA*: who gets the protection afforded to customers of water and wastewater services when the property to which the service is supplied is the subject of a lease? In New South Wales, s 17 of the *Retail Tenancies Act* 1987 (NSW) presently stipulates the allocation of charges between lessor and lessee. The lessor must pay the rates, taxes and charges specified under the Act, other than charges for electricity, gas, excess water and other prescribed charges. Changes are being mooted that would pass all water charges on to tenants, not just excess water charges (OFT, *A New Direction* 2007). The present scheme means that the lessor must pay the base or fixed water charge and hence it is the lessor who contracts with the water- and wastewater-service provider. Further, when determining prices, IPART allocates costs between service charges — for which the lessor is liable — and usage charges — for which the lessee is liable (PIAC, *Creating a Dynamic and Competitive Water Industry* 2006: 16). Under the access regime, third-party providers will not have their prices regulated by IPART (unlike the public utility providers). They will be determined by the market. New providers will, therefore, need to undercut the IPART prices or, alternatively, charge more but lure customers by the provision of better or increased services. New providers will not be subject to IPART's scrutiny about cost allocation between landlord and tenant and this may mean that lessors enter into contractual arrangements with new providers that shift costs from lessor to lessee (PIAC, *Creating a Dynamic and Competitive Water Industry*: 16).

Conclusion

This chapter has provided an overview of the present institutional frameworks for water and wastewater management throughout Australia with the aim of identifying where opportunities might exist for third parties to compete with public utilities for service provision.

This was followed by an examination of the legal frameworks under which applications may be brought for access to monopoly infrastructure essential for competing in the upstream and downstream markets for service provision. The first considered was the Commonwealth legislation, contained in Part IIIA of the *Trade Practices Act* 1974, on which Services Sydney relied.

The *Water Industry Competition Act* 2006 (NSW) was then examined as the first example of state-based legislation designed, in part at least, to facilitate competition through the provision of third-party access to infrastructure services. The chapter observed the importance of *WICA* in its potential to serve as a model for other states' legislation.

The final part of the chapter provided a discussion of some of the key issues associated with third-party access to infrastructure services. They included: coverage declarations; access and resource pricing; building customer trust, step-in rights; property rights; negotiating inexperience; and leases.

What emerged is that legal regimes and frameworks are capable of being designed to facilitate third-party access by providing frameworks for the negotiation and arbitration of the terms of access to the essential infrastructure service, along with appropriate health- and consumer-protection safeguards, for example. However, it is possible that such regimes will still fail, often for reasons beyond the law. Third-party access regimes in the sewerage sector may flounder because of access-pricing issues, a lack of confidence in new entrants and market shrinkage due to shifts in favour of home- or community-based recycling schemes, for example. The legal problems that potentially exist in relation to third-party access, such as ownership of the raw resource, would seem capable of being resolved. Some of the political, economic, social and cultural problems would appear to present greater difficulties.

The third-party access regimes may also fail in relation to sewer mining for a lack of coverage. If sewage appreciates in resource value, the sewerage-service providers, as potential 'owners' of the sewage, will have considerable power in allocating the resource to the highest bidder, especially if the regulatory authorities do not or cannot regulate the price of sewage to the sewer miners. In NSW, the current legislation authorises IPART to regulate pricing by governmental water agencies, but what will be the position for private sewerage-service providers? It is possible that there is a legislative gap, deliberately left because of notions of private property pertaining to ownership of sewage. Will governments and parliaments be forced, at some stage in the future, to legislate for a resource-regulation regime based on public ownership of sewage and governmental allocation of this fantastic new resource?

Although it is legally possible to 'unbundle' various aspects of the wastewater sector (for example, by separating out extraction, treatment, distribution, household connection, billing, maintenance and construction of infrastructure) in order to create spaces for competitive third-party involvement, problems may still emerge. For example, if obligation is broken up and shared along the supply chain, it may be easier to avoid responsibility for system failures. If there is a problem, it is always potentially the fault of someone else. Bearing in mind Godden's warning (in Chapter 8, following) that '[m]oves to deregulate urban water authorities have created models of governance that transcend the simplistic view of dichotomous public and private spheres and public/private property', it is perhaps still possible that David Hare captured some of the difficulties associated with the unbundling process in his play the *Permanent Way*, which dealt with the privatisation of British Rail. One of his characters observed:

> Everyone knows: the Balkanisation was a complete disaster. The thing was broken up into 113 pieces, like beads thrown onto a table, all to be held together by local contracts and all in pursuit of the idea of competition. Well, competition on the railways is a great idea in the theory, hopeless in practice. (Hare 2003: 18)

On this analysis, an unbundled water sector may lead to enhanced competition, which, of course, is the aim of *WICA*, but whether that competition, in turn, leads to better outcomes for society, the environment and consumers is more problematic.

References

Australian Government, Department of the Prime Minister and Cabinet (AGD of PMC) 2006, 'A Discussion Paper on the Role of the Private Sector in the Supply of Water and Wastewater Services', available via link at: http://www.pmc.gov.au/water_reform/index.cfm#urban_water

Bakker, K. 2003, *An Uncooperative Commodity: Privatising Water in England and Wales*, OUP: Oxford.

Bentham, J. *Principles of the Civil Code*, C. K. Ogden (ed.) 1931; Chapter viii cited in C. B. Macpherson 1978 (ed.), *Property: Mainstream and Critical Positions*, Blackwell: Oxford.

Bonyhady, T. 2000, *The Colonial Earth*, Melbourne University Press: Melbourne.

'Competition Principles Agreement', available via link at: http://www.ncc.gov.au/publication.asp?activityID=39&publicationID=99

Department of Environment, Food and Rural Affairs (DEFRA) 2005, *Water Act 2003 — Water supply licensing: Response to the Commodification on Collective Modifications of Standard Licence Conditions of Water Supply Licences*: 1.

Department of Water and Energy (DWE) 2007, 'Water Industry Competition Act 2006, Regulations Consultation Paper', available at: http://www.waterforlife.nsw.gov.au/__data/assets/pdf_file/0008/7892/190615_WICA_Consultation_paper.pdf

Dingle, A. 2007, 'Reflections on the history of water supply and sewerage provision in Australia's cities', ASSA Annual Symposium, 'Power, People, Water: Urban Water Services and Human Behaviour in Australia', Canberra, 20 November.

Dovers, S. 2007, 'Urban water: policy, institutions and governance', ibid.

Economic Regulation Authority (ERA) Western Australia 2007, 'Issues Paper, Inquiry on Competition in the Water and Wastewater Services Sector', 20 July, available at: http://www.era.wa.gov.au/cp-root/5765/26496/20070720%20Issues%20Paper%20-%20Inquiry%20on%20Competition%20in%20the%20Water%20and%20Wastewater%20Services%20Sector.pdf

Economic Regulation Authority, Western Australia, 'Inquiry on Competition in the Water and Wastewater Services Sector', Draft Report, 3 December 2007.

Fisher, D. E. 2000, *Water Law*, LBC, Sydney.

Freehills Solicitors, 'Competition and Market Regulation Update', November 2006, available at: http://www.freehills.com.au/publications/publications_6730.asp

Godden, L., *Property in Urban Water: Private Rights and Public Governance*, ASSA Annual Symposium op. cit.

Gray, J. 2008, 'Watered Down? Legal Constructs, Tradeable Entitlements and the Regulation of Water' in D. Ghosh, H. Goodall and S. Donald (eds), *Sovereignty and Borders in Asia and Oceania*, Routledge, forthcoming.

Gray, K., 1991, 'Property in Thin Air', *CLJ 252*.

Gray, K. and Gray, S. 2004, *Elements of Land Law*, Oxford University Press: Oxford.

Hare, D. 2003, *The Permanent Way*, Faber and Faber: London.

Hawkins, G. 2006, *The Ethics of Waste: How we Relate to Rubbish*, UNSW Press: Sydney.

Independent Pricing and Regulatory Tribunal (IPART) 2005, 'Investigation into Water and Wastewater Service Provision in the Greater Sydney Region — Final Report', (IPART Water and Wastewater Report), October. Available at: http://www.ipart.nsw.gov.au/files/Section%209%20Investigation%20into%20Water%20and%20Wastewater%20Service%20Provision%20-%20Final%20Report%20-%2028%20November%202005.PDF

Independent Pricing and Regulatory Tribunal (IPART) 2006, *Pricing Arrangements for Recycled Water and Sewer Mining*, September. Available at: http://www.ipart.nsw.gov.au/files/Final%20Report%20-%20Pricing%20Arrangements%20for%20recycled%20water%20and%20sewer%20mining%20-%20September%202006%20-%20website%20document.PDF

Independent Pricing and Regulatory Tribunal (IPART) 2007, *Promoting Retail Investment and Competition in the New South Wales Electricity Industry: Regulated Electricity Tariffs and Charges for Small Customers 2007–2010*, Electricity: Final Report and Final Determination, June. Available via link at: http://www.ipart.nsw.gov.au/investigation_content.asp?industry=2&sector=3&inquiry=108&doctype=7&doccategory=1&docgroup=1

Independent Pricing and Regulatory Tribunal (IPART), *Overview of Final Report and Determination on Electricity Retail Prices in New South Wales from 1 July 2007– 30 June 2010, Promoting Retail Investment and Competition in the New South Wales Electricity Industry: Regulated Electricity Tariffs and Charges for Small Customers*, June 2007. Available at: http://www.ipart.nsw.gov.au/files/Fact%20Sheet%20-%20Overview%20of%20Final%20report%20and%20determination%20on%20electricity%20retail%20prices%20in%20NSW%20-%201%20July%202007%20to%2030%20June%202010.PDF

Marsden Jacob and Associates, Research Paper prepared for the Australian Government Department of Agriculture, Fisheries and Forestry, 'Third-party Access in Water and Sewerage Infrastructure: Implications for Australia', 22 December 2005. Available at: http://www.daffa.gov.au/__data/assets/pdf_file/0019/29260/3_water_sew_infr.pdf

Metropolitan Water Directorate, the Cabinet Office, NSW (MWDCO), 'Consultation Paper: Creating a Dynamic and Competitive Water Industry', May 2006.

National Competition Council (NCC Principles for Reform) Occasional Series, *Principles for National Reform, Learning Lessons from the NCP*, October 2005. Available at: http://www.ncc.gov.au/pdf/PIReNCP-003.pdf.

National Competition Council's Draft Recommendations on the application by Services Sydney for Declaration of Sewage Transmission and Interconnection Services Provided by Sydney Water Corporation (NCC Draft Recommendations), November 2004. Available at: http://www.ncc.gov.au/pdf/DEWASSSu-022.pdf

New South Wales Cabinet Office (NSWCO), *Water Industry Competition (Access to Infrastructure Services) Regulation 2007*. Available at: http://www.cabinet.nsw.gov.au/__data/assets/pdf_file/0012/8400/Draft_Regulation.pdf

Office of Fair Trading (OFA) Consultation Paper entitled 'Residential Tenancy Law Reform, A New Direction', released 20 September 2007. Available at: http://www.fairtrading.nsw.gov.au/pdfs/realestaterenting/residentialtenancylawreformanewdirection.pdf

Public Interest Advocacy Centre (PIAC), *Submission to Consultation Paper: Creating a dynamic and competitive metropolitan water industry*, 7 June 2006. Available at: <http://www.waterforlife.nsw.gov.au/__data/assets/pdf_file/0007/1501/PublicInterestAdvocacyCentre_070606.pdf > Accessed 12 September 2007.

Queensland Water Commission (QWC), *Our Water: Urban water supply arrangements in South East Queensland, Final Report*, May 2007. Available via link at: http://www.qwc.qld.gov.au/Urban+water+supply+arrangements

Queensland Water Commission (QWC) *Fact Sheet 1, Urban Water Supply Arrangements in SEQ: Changes from Present Arrangements*. Available via link at: http://www.qwc.qld.gov.au/Urban+water+supply+arrangements

Slater, S. 2005, 'A Prescription for Fulfilling the Promise of a Robust Water Market', *McGeorge Law Review* 36: 254–93.

Spearritt, P., *Water Crisis in South East Queensland*, ASSA Annual Symposium, op. cit.

Sydney Water Corporation, 'Indicative Network Access Agreement', Discussion Draft, 1 August 2007.

Sydney Water, *Sewer Mining: How to establish a sewer mining operation*, May 2006, available at: <http://www.sydneywater.com.au/Publications/FactSheets/SewerMiningHowToEstablishASewerMiningOperation.pdf#Page=1>

Cases and determinations

Application by Services Sydney Pty Ltd [2005] ACompT2 (11 April 2005) at [201]. Available at: http://www.austlii.edu.au/au/cases/cth/ACompT/2005/7.html

Australian Competition and Consumer Commission, Access to Dispute between Services Sydney Pty Ltd and Sydney Water Corporation, Arbitration Report, 19 July 2007. Available at: http://www.accc.gov.au/content/item.phtml?itemId=793015&nodeId=42c60919002f38d8ea73088b1fbeda82&fn=Arbitration%20Report%20Final%20July%202007.pdf.

Doodeward v Spence [1908] HCA 45; (1908) 6 CLR 406 (31 July 1908).

Drake v Minister for Planning [2003] NSWLEC 270.

Moore v The Regents of the University of California (1990) 793 P 2d 479.

R v Herbert [1961] JPLGR 12.

R v Welsh [1974] RTR 478.

Venner v State of Maryland 354 A2d 483 (1976)

Legislation and Codes of Practice.

Australian Drinking Water Guidelines and the *NSW Plumbing Code of Practice* 2006

Environmental Planning and Assessment Act 1979 (NSW)

Local Government Act 1993 (NSW)

Public Health Act 1991 (NSW)

Protection of the Environment Operations Act 1997 (NSW)

Sydney Water Act 1994 (NSW)

Retail Tenancies Act 1987 (NSW)

Trade Practices Act 1974 (Cth)

Water Industry Competition Act 2006 (NSW)

Water Management Act 2000 (NSW)

ENDNOTES

[1] This poem is part of a collection of verses entitled *The Man from Snowy River,* which was reinterpreted in a film of the same name.

[2] See T. Bonyhady, *The Colonial Earth*, Melbourne University Press, Melbourne: 2000.

[3] For a broad discussion of waste, see G. Hawkins, *The Ethics of Waste: How we Relate to Rubbish*, UNSW Press, Sydney: 2006.

[4] The Bunbury Water Board, which trades as AQWEST, provides potable water to the Bunbury-Wellington Region. AQWEST does not provide wastewater services. Busselton Water Board, trading as Busselton Water, is another statutory authority providing similar services to AQWEST.

[5] Apparently, 90 per cent of the Water Corporation's capital projects were put out to tender in 2006. See *Issues Paper, Inquiry on Competition in the Water and Wastewater Services Sector*, Economic Regulation Authority, Western Australia, 20 July 2007, p.2. Available at: http://www.era.wa.gov.au/cp-root/5765/26496/20070720%20Issues%20Paper%20-%20Inquiry%20on%20Competition%20in%20the%20Water%20and%20Wastewater%20Services%20Sector.pdf

[6] United Water is 95 per cent owned by Veolia Water Australia.

[7] As at July 2007, a review was in progress which would consider the amalgamation of councils responsible for providing water and wastewater services. See *Issues Paper,* op. cit. note 5.

[8] These parasites can cause gastroenteritis in humans and as a result of unacceptably high levels being found in their drinking water Sydneysiders had to boil all drinking water while the problem lasted. See Sydney Water Inquiry (New South Wales Premier's Department, 1998). According to C. Sheil, (*Water's Fall: Running the Risks with Economic Rationalism*, Pluto Press, Sydney: 2000), McLellan noted that water from the privatised Prospect treatment plant was probably a source of the problem; but the company (Australian Water Services) which built and operated the plant denied all responsibility because its contract did not require that it test for Cryptosporidium or Giardia. The report itself noted two probable reasons for the presence of the Cryptosporidium and Giardia: heavy rains following a period of drought which transported the organisms from the catchment to the supply infrastructure, and operational difficulties at the main treatment plant. See also A. Gardner 2003, 'Law and Policy for Sustainable Water Quality Management: Focus on the Sydney Water Catchments', *The Australasian Journal of Natural Resources Law and Policy* 8: 99–128.

[9] In 2007, 107 local and county councils provide water, sewerage and drainage services in rural and regional New South Wales. See Department of Water and Energy, *Water Industry Competition Act, 2006 Regulations and Consultation Paper*, June 2007. Available at: http://thecabinetoffice.clients.squiz.net/__data/assets/pdf_file/0008/7892/190615_WICA_Consultation_paper.pdf

[10] There are five smaller public water-utilities supply authorities that operate water, supply, sewerage and or drainage function under the *Water Management Act* 2000 (NSW). These include Gosford City Council and Wyong Shire Council.

[11] The third-party access regime is restricted to facilitating access to the excess capacity of the infrastructure in that existing user rights may not be disturbed without compensation: see *Trade Practices Act* 1974 (Cth) s 44W.

[12] *BHP Billiton Iron Ore v The National Competition Council* [2007] FCAFC 157, per Greenwood J at [125] explaining the purpose of Part IIIA of the *TPA*. See also *Water Industry Competition Act* 2006 (NSW) s 21.

[13] Note that the term 'third-party access' covers 'access to essential facilities', 'common carriage' and 'mandatory unbundling'. Sewage sludge (the solids) may be disposed of in a variety of ways. For example,

see the work of Dr Sandra Maria Campos Alves — CSIRO Land and Water, the University of Adelaide, whose PhD thesis deals with sewage sludge application in corn cultivations. In Galway, Ireland, the sludge is thickened and pasteurised and the dried sludge is dewatered and transported to sites where it is spread on land as an organic fertiliser. See also *Issues Paper, Inquiry on Competition in the Water and Wastewater Services Sector*, op. cit.

[14] See also *Application by Services Sydney Pty Limited* [2005] ACompT 7 at [45] and IPART's description of sewer mining; *Pricing Arrangements for Recycled Water and Sewer Mining*, September 2006: 65.

[15] The authors wish to acknowledge the assistance of Dr Wendy Timms, Senior Engineer, Water Research Laboratory, UNSW School of Civil and Environmental Engineering, Manly Vale, NSW for her assistance in explaining the practicalities of sewer mining. Note that sewer miners in the Sydney region need to gain local council approval for the particular end-use of the recycled water.

[16] Sewer-mining projects in Sydney include those at Kogarah Golf Course and the 'Olympic' suburb of Newington. Other such operations exist in Southwell Park in the ACT (which has been operating since 1995), Cranbourne Sewer Mining Facility near Melbourne (which has been operating since 1974), the Kings Domain Gardens in Melbourne and the Rocks Water Mining Project in Brisbane.

[17] The draft agreement envisages comprehensive monitoring of the third-party access seeker's customer-input volumes. It also acknowledges that a certain level of flow is necessary for the sewerage system to operate effectively while, at the same time, the third-party access seeker needs to retain a certain volume of sewage to operate its plant. Accordingly, an interdependent Flow Management Agreement is contemplated in cl 8.1(b)(iv).

[18] See Davison, Chapter 3 of this volume, for an historical discussion of the importance of piped and flowing water to sewerage systems.

[19] The Australian Energy Market Commission sets rules and policy in the electricity sector. It operates in conjunction with the National Electricity Law, which is contained in a Schedule to *National Electricity (South Australia) Act* 1996. The NEL is applied as law in each participating jurisdiction of the NEM by application statutes; for example, the *National Electricity (Victoria) Act* 2005. In turn, the legislation is supported by regulation providing access regimes. Observe that less frequently third-party access has applied to transport supply chains such as sea ports and airports. See Marsden and Associates, Research Paper prepared for the Australian Government Department of Agriculture, Fisheries and Forestry, *Third-party Access in Water and Sewerage Infrastructure: Implications for Australia*, 22 December 2005: 9. Available at: http://www.daffa.gov.au/__data/assets/pdf_file/0019/29260/3_water_sew_infr.pdf> See also L. Flor and E. Defilippi 2003, 'Port Infrastructure: An Access Model for the Essential Facility', *Maritime Economics and Logistics*, 5 (2): 116–32.

[20] Economic Regulation Authority, Western Australia, *Inquiry on Competition in the Water and Wastewater Services Sector,* Draft Report, 3 December 2007 p.(*v*) notes that at least in the short term, the uncertainty of supply makes the water sector different from other utility industries such as gas and electricity. Available via link at: http://www.era.wa.gov.au/2/508/46/inquiry_into_co.pm

[21] Department of Environment, Food and Rural Affairs (DEFRA), *Water Act 2003 — Water supply licensing: Response to the Consultation on Collective Modifications of Standard Licence Conditions of Water Supply Licences*, June: 1. For further discussion, see Marsden and Associates, Research Paper prepared for the Australian Government Department of Agriculture, Fisheries and Forestry, *Third-party Access in Water and Sewerage Infrastructure: Implications for Australia*, 22 December 2005: 17. Available at: http://www.daffa.gov.au/__data/assets/pdf_file/0019/29260/3_water_sew_infr.pdf

[22] Ofwat is the water and sewerage-service regulation authority in England and Wales.

[23] Litigation has ensued. See *Metro, Water Dist.*, 96 Cal Rptr.2d at 316; *San Luis Coastal Unified School Dist. v City Morro Bay* S. B. 621 (2002) (as amended in the Assembly on 28 August 2002).

[24] Marsden Jacob and Associates, op. cit. (note 19)

[25] In October 2004, Lakes R Us applied to the NCC under Part IIIA of the *TPA*, 1974. The NCC recommended to the Premier that no declaration be made. Eventually Lakes R Us decided not to pursue its application.

[26] Note that the Economic Regulation Authority, Western Australia, *Inquiry on Competition in the Water and Wastewater Services Sector*, Draft Report, 3 December 2007, p. *xi*, recommends that a state-based third-party access regime be implemented in Western Australia.

[27] Officially known as the 'National Competition Policy Report'.

[28] The introduction of the *Water Management Act* 2000 (NSW) is an example of legislative change initially tied to NCC payments while the *Water Industry Competition Act* 2006 (NSW) follows up on the government's commitment to private-sector involvement found in Metropolitan Water Directorate, *Consultation*

Paper: Creating a dynamic and competitive metropolitan water industry, May 2006: 4. Available at: http://www.waterforlife.nsw.gov.au/__data/assets/pdf_file/0013/1444/smwp_consult_paper_1.pdf

See R. Lyster, Z. Lipman, N. Franklin, G. Wiffer and L. Pearson, *Environmental and Planning Law in New South Wales*, Federation Press, Sydney, 2007: 275.

[29] *TPA* s 44N provides for the Commonwealth Minister to declare that a State regime is an effective access regime.

[30] If more than one state provides the service for which access is sought, vested jurisdiction or other co-operative legislative schemes should be used so that parties need only deal with a single access process. See *Competition Principles Agreement*, April 1995, s 6 (4), a copy of which is available in National Competition Council, *Compendium of National Competition Policy Agreements*, 2nd edition, June 1998, a vailable via link at: http://www.ncc.gov.au/articleZone.asp?articleZoneID=16.

[31] The criteria are set out in *TPA* Part IIIA, Division 2, ss 44G(2) for the Council and 44H(4) for the Minister.

[32] *TPA* s 44B; where "service" is defined for the purposes of Part IIIA.

[33] This definition is currently being litigated in the context of third-party access to a railway line used for the transport of iron ore in Western Australia's Pilbara region: *BHP Billiton Iron Ore v The National Competition Council* [2007] FCAFC 157. The issue there concerns the paragraph (f) exclusion of a production process from the definition of a service. The Full Federal Court held that the use of the rail line as part of a production process does not qualify for the exemption. The opposite conclusion could have significant implications for a similar argument by a sewerage-service provider that its sewerage network was part of a production process for recycled water.

[34] *Application by Services Sydney Pty Limited* [2005] ACompT 7.

[35] Postage-stamp pricing is uniform pricing, meaning that the same price per unit is charged irrespective of where the sewage is transported from across Sydney.

[36] The ACCC ruling did not determine a pricing formula for the interconnection services, because that issue was not ripe for determination.

[37] According to Andrew Stoner, Member for Oxley, Leader of the National Party, Legislative Assembly Hansard (Extract) 14 November 2006 (available at: http://www.parliament.nsw.gov.au/prod/PARLMENT/hansArt.nsf/V3Key/LA20061114036>), the New South Wales government spent approximately, $1.6 million in legal fees opposing the application.

[38] For an analysis of the lack of economic viability, see National Competition Council's Draft Recommendations on the application by Services Sydney for Declaration of Sewage Transmission and Interconnection Services Provided by Sydney Water Corporation, November 2004: 30–1. Available at: http://www.ncc.gov.au/pdf/DEWASSSu-022.pdf. A 'dependent market' is the defined market for the proposed services.

[39] The pricing methodology was set out in the Australian Competition and Consumer Commission's pricing determination of July 2007. See *Access Dispute between Services Sydney Pty Ltd and Sydney Water Corporation, Arbitration Report,* 19 July 2007, available via link at: http://www.accc.gov.au/content/index.phtml/itemId/793015

[40] *WICA* received assent on 27 November 2006 but will only commence once the Regulations have been drafted and finalised.

[41] For example, Western Australia, where the Economic Regulation Authority is currently conducting an Inquiry on Competition in the Water and Wastewater Services Sector: see http://www.era.wa.gov.au/

[42] http://www.waterforlife.nsw.gov.au/__data/assets/pdf_file/0015/1446/wfl2006.pdf at p.115.

[43] IPART, 'Investigation into Water and Wastewater Service Provision in the Greater Sydney Region-Final Report', October 2005 available at: http://www.ipart.nsw.gov.au/files/Section%209%20Investigation%20into%20Water%20and%20Wastewater%20Service%20Provision%20-%20Final%20Report%20-%2028%20November%202005.PDF. See also Metropolitan Water Directorate, *Consultation Paper: Creating a dynamic and competitive metropolitan water industry*, May 2006: 4. Available at: http://www.waterforlife.nsw.gov.au/__data/assets/pdf_file/0013/1444/smwp_consult_paper_1.pdf

[44] See long title of Act available at: http://www.austlii.edu.au/au/legis/nsw/consol_act/wica2006333/longtitle.html

[45] Regulations for access are being developed in a separate process.

[46] *WICA* 2006 s 21, statement of the objects of Part III, Access to infrastructure services.

[47] See Indicative Sewer Network Access Agreement between Sydney Water Corporation and the Access Rights Holder, Discussion Draft, 1 August 2007. Under cl 8(2)(a) Sydney Water agrees to make available the Extraction Proportion/Volume for extraction by the Access Rights Holder while cl 8(2) (b) requires the Access Rights Holder to extract the agreed volume or proportion. It is possible, however, that 'ownership' of the sewage will first need to be clarified and resolved.

[48] *WICA* (NSW) s 23 (a)–(e) or if the access is declared from the outset s 22 and Schedule 1.

[49] Arbitration is governed by *WICA* (NSW) s 40, in conjunction with the Regulations and the *Commercial Arbitration Act* 1984, as well as s 24B–24E of the *Independent Pricing and Tribunal Regulatory Act* 1992. IPART must publish the arbitration determination on its website. See *WICA* (NSW) s 40 (11).

[50] Refer to earlier section in this paper 'Current Institutional Frameworks for Water and Wastewater'.

[51] It is notable that in August 2007, SA Water and United Water were heavily criticised in a report based on a two-year study undertaken by the Total Environment Centre. It found that Adelaide's water-pricing scheme was the worst in the country and that the two above-mentioned companies were lacking in transparency and accountability. Adelaide's water-pricing process was also regarded as the 'worst in the nation'. See http://www.abc.net.au/news/stories/2007/08/10/2001605.htm. This was not the first time that the private operator, United Water, had been the subject of criticism. In 1996, one year after the United Water contract was signed, the company was also criticised for its part in what was colloquially known as Adelaide's 'big poo', when a 'widespread and offensive stench' covered the city. A team of Queensland University investigators ultimately traced the smell to raw sewage, which had accidentally been released into open treatment ponds at the city's Bolivar treatment plant. The lead investigator was quoted as concluding: 'It was dollars driving everything. The big emphasis was on minimising costs. The Bolivar incident is an illustration of what can happen when things like monitoring and maintenance are cut to the bone.' See P. Mac, 'Water Privatisation the problem, not the solution', *The Guardian*, 14 May 2003, Issue No 1137 available at: http://www.cpa.org.au/garchve03/1137water.html

[52] The authors also wish to acknowledge their indebtedness to Luke Woodward, Partner, Gilbert + Tobin Lawyers, Sydney, for his discussions in relation to the *WICA* Licensing Scheme Regulations.

[53] NSW Cabinet Office, *Water Industry Competition (Access to Infrastructure Services) Regulation 2007* available at: http://www.cabinet.nsw.gov.au/__data/assets/pdf_file/0012/8400/Draft_Regulation.pdf

[54] NSW Department of Premier and Cabinet, Technical assistance for the Development of the Water Industry Competition (Access Regulation) Final Report, May 2007: 2. Available at: http://www.cabinet.nsw.gov.au/__data/assets/pdf_file/0011/8399/GHD_Report.pdf

[55] This concept is explained by R. Rhodes, 'The Hollowing Out of the State', 65 *Political Quarterly* 138 (1964).

[56] See N. Gunningham and P Grabosky, *Smart Regulation: Designing Environmental Policy*, Clarendon Press, Oxford: 1988; Australian Government, Office of Best Practice Regulation, *Best Practice Regulation Handbook*: November 2006; Australian Government, *Rethinking Regulation, Report of the Taskforce on Reducing Regulation Burdens on Business*, January 2006. Available via link at: http://www.treasury.gov.au/contentitem.asp?ContentID=1141&NavID=

[57] J. Black, 'Tensions in the Regulatory State' 2007, *Public Law* 58–73: 62. See also J. Black 2002, 'Critical Reflections on Regulation' 27 *Australian Journal of Legal Philosophy* 1; and CARR Discussion Paper (2002) available at: www.lse.ac.uk/collections/carr

[58] Andrew Stoner, Member for Oxley, Leader of the National Party, Legislative Assembly Hansard (Extract), 14 November 2006. Available at: http://www.parliament.nsw.gov.au/prod/PARL-MENT/hansArt.nsf/V3Key/LA20061114036

[59] See *Application by Services Sydney Pty Limited* [2005] ACompT 7 at [192]–[208], where the Tribunal discusses mainly arguments about the public interest in providing access under the then proposed state regime and potentially adverse impacts on service pricing.

[60] Personal communication from Dr Michael Keating, Chair of IPART to Alex Gardner, November 2007.

[61] *WICA* s 41(3). 'Postage-stamp pricing' is defined in s 41(3) as the system of pricing in which the same kinds of customers within the same area of operations are charged the same price for the same service.

[62] Based on private conversations between water law practitioners and Janice Gray on 7 August 2007.

[63] Andrew Stoner, Member for Oxley, Leader of the National Party, New South Wales Legislative Assembly, Hansard (Extract), 14 November 2006. Available at: http://www.parliament.nsw.gov.au/prod/PARLMENT/hansArt.nsf/V3Key/LA20061114036 .

[64] For a list of independent reasons why opportunities for third-party access may not be taken up, see Marsden Jacob Associates, op. cit.: 86.

[65] See following discussion of property rights.

[66] See Department of Water and Energy, *Water Industry Competition Act 2006, Regulations Consultation Paper,* June 2007: 33. Available at: http://www.waterforlife.nsw.gov.au/__data/assets/pdf_file/0008/7892/190615_WICA_Consultation_paper.pdf

[67] Public Interest Advocacy Centre, *Submission to Consultation Paper: Creating a dynamic and competitive metropolitan water industry*, 7 June 2006: 5. Figures said to be based, in part, on research commissioned by IPART.

[68] Ibid.

[69] The circumstances included: (1) the high levels of dissatisfaction with the incumbent electricity supplier in South Australia; (2) the South Australian Government's offer to concession-card holders of a $50 cash rebate to switch; (3) the existence of comparative price information services in Victoria and South Australia which do not exist in New South Wales; and (4) the fact that Victoria and South Australia are the second- and third-most active markets in the world. See IPART, *Promoting Retail Competition and Investment in the NSW Electricity Industry: Regulated Electricity Tariffs and Charges for Small Customers 2007–2010.* Available at: <http://www.ipart.nsw.gov.au/investigation_content.asp?industry=2&sector=3&inquiry=108&doctype=7&doccategory=1&docgroup=1<

[70] See http://www.abc.net.au/news/stories/2007/12/08/2113412.htm; for popular comment see: http://www.crikey.com.au/Business/20071206-NSW-Electricity-privatisation-will-Luddite-economics-rule-in-NSW.html

[71] The Federal Minister for Communications implemented the 'Do Not Call Register' on 3 May 2007. The Register lists the telephone numbers of people and organisations that telemarketers are not to call.

[72] Retail headroom involves making it easier for new retailers to enter the market and win customers, by raising the regulated price of water for consumers, above that which reflects efficient costs.

[73] This scenario also assumes that the public utility staff who were retrenched either took positions outside the industry or lost their skills in some way.

[74] See reference to Adelaide's 'big pong' under section '*WICA*'s relationship with other legislation'.

[75] Sydney Water's draft third-party access agreement asserts, in the recitals, its ownership of infrastructure. Draft agreement made available to Janice Gray, 17 September 2007.

[76] See D. E. Fisher, *Water Law*, LBC, Sydney, 2000; J. Gray, 'Watered Down? Legal Constructs, Tradeable Entitlements and the Regulation of Water' in D. Ghosh, H. Goodall and S. Donald (eds) 2008, *Sovereignty and Borders in Asia and Oceania*, Routledge, forthcoming.

[77] For example, if I collect urine to take to a pathologist for testing, the urine is most likely to be my property. (See *R v Welsh* [1974] RTR 478). If my hair is cut, the cut hair is my property to sell to a wig-maker, for example. (See *R v Herbert* [1961] JPLGR 12). See also *Venner v State of Maryland* 354 A2d 483 (1976) at 498–9 on the question of human waste. Note human bodily waste is literally created by the householder. It is certainly property after it has been treated as sludge and it is possibly property beforehand; but see *Moore v The Regents of the University of California* (1990) 793 P 2d 479. In that case the court had to decide on whether it was possible to have property rights in human tissue; that is, to be the owner of a cell line. In a split decision, it found that there was no property in the cell line but the decision in this case was influenced by the fact that legislation had taken property rights away from human body parts. See also *Doodeward v Spence* [1908] HCA 45; (1908) 6 CLR 406 (31 July 1908) per Griffith CJ for a discussion of property in the human body after death. Note also that even if the human bodily waste component of sewage is not property, property may still exist in the newly created product of sewage. To explain, although air itself might not be the subject of property, a sponge cake that is aerated is no less property because it contains air. Property exists in the new form of a cake.

[78] For example, Sydney Water Customer Contract — see http://www.sydneywater.com.au/WhoWeAre/OperatingLicence/CustomerContract.cfm. See also the *Sydney Water Act* 1994 (NSW).

[79] Note that although the language of 'control' is used in relation to sewage, IPART does not go as far as saying that incumbent water agencies actually have property in the sewage. See IPART, *Pricing Arrangements for Recycled Water and Sewer Mining*, September 2006, p.66. Available at: http://www.ipart.nsw.gov.au/files/Final%20Report%20-%20Pricing%20Arrangements%20for%20recycled%20water%20and%20sewer%20mining%20-%20September%202006%20-%20website%20document.PDF

[80] See *Haynes' Case* (1614) 12 Co Rep 113, 77 ER 1389. *R v Edwards and Stacey* (1877) 13 Cox CC 384 at 385; *Tancil v Seaton* (1877) 26 AmRep 380 at 382, 28 Gratt 601; *Vincent State Bank of NSW Ltd* (Unreported, Supreme Court of NSW, 30 July 1993). Accordingly, K. Gray and S. Gray, *Elements of Land Law*, Oxford University Press, Oxford, 2004: 55, point out that one cannot abandon a chattel on one's own land.

[81] Argued by parity of reasoning with *Munday v Australian Capital Territory* (Unreported, Supreme Court, ACT, 8 July 1998) per Higgins J.

[82] See *R (Thames Water Utilities) v South East London Division, Bromley Magistrates' Court* (Environment Agency as interested party), Court of Justice of the European Communities, Second Chamber, Case C-252/05, 10 May 2007. (Environment Agency as an interested party), heard by the Second Chamber of the Court of Justice of the European Communities, on a reference for a preliminary ruling from the Queen's Bench Division (Administrative Court).

[83] In the Australian context, see *Drake v Minister for Planning* [2003] NSWLEC 270 for a discussion of (a) waste as something valuable and (b) 'goods'. Also see *Protection of the Environment Operations Act 1997* (NSW).

[84] K. Gray 1991, 'Property in Thin Air', *CLJ 252*. Note, in a strata scheme there is property in a horizontal airspace.

Chapter 8

Property in urban water: Private rights and public governance

Lee Godden

> I love a sunburnt country
> A land of sweeping plains
> Of rugged mountain ranges,
> Of droughts and flooding rains.
> I love her far horizons
> I love her jewel sea
> Her beauty and her terror
> The wide brown land for me.
>
> …
>
> For flood and fire and famine
> She pays us back threefold …
>
> (Dorothea Mackellar, *My Country*)

How ironic must the words of Dorothea Mackellar now seem in an era of climate change, drought, and increasing pressure on water across all facets of Australian life? To the first stanza of this poem, duly mouthed by the sing-song voices of generations of Australian children, I have added two lesser-known lines from a later stanza. What I would suggest, moreover, is that, in fact, the debt runs the other way — the Australian people have taken threefold and our use of water runs at highly unsustainable levels in both rural and urban areas. The debt will not be paid so much by current water users, but by future generations who will endure either wide-ranging environmental degradation and/or long-term taxpayer-funded repayments. This recognition can be something of a truism allowing a predominately 'business as usual' approach to managing and regulating water in urban contexts in Australia, or else it can be the catalyst for a more fundamental rethinking of how water is conceptualised, valued, managed and allocated across the Australian continent. The manner in which water is regulated reflects not only broader social values but also that its governance occurs in relation to the particular legal constructs that define rights and responsibilities for water. Accordingly, property concepts provide a key means for articulating the choices that arise as Australian society, through its institutional and governance frameworks, responds to the urban 'water crisis'.

Property, though, far from being the settled and determinate concept that people ascribe to the word, remains a contested site for defining 'rights' (Fisher 2007). Crucially also, if we are to begin a more fundamental re-evaluation of how water is integral to living in an increasingly urban society, property must also be regarded as a site for more clearly articulating obligations and responsibilities in respect of long-term sustainability and inter-generational equity. Indeed, whether property, with its lay connotations of privatisation and individual rights, is an appropriate descriptor for defining and managing 'entitlements' to water in an urban context where 'water security' is becoming a pervasive discourse (Marsden Jacob 2006) needs to be closely examined. Thus, this paper explores the multifarious ways in which 'property' and 'rights' in water' might be understood in the urban context. Further, it examines the role that property might play in resolving crucial tensions related to the balance between private forms of ordering relationships to water and the dynamic of the public interest. In particular, it explores how trends to deregulate state-based institutional models of water governance operate in concert with moves to institute 'water-security projects'. In this manner, it examines the ramifications of these emerging governance and institutional forms such as water trade, and market mechanisms for the relationships being articulated between urban and rural areas within Australia. To illustrate this dynamic interplay, the paper examines the development of water infrastructure as an interface between urban and rural areas by reference to the Food Bowl Modernisation Project in Victoria and concurrent metropolitan water-industry reforms. This examination suggests that there is not a neat division between public and private spheres but that property in water is articulated against the complexities of water 'resource development' and water regulation.

Restructuring the public/private spheres of water regulation

'The prolonged process of social and economic restructuring of the relations between the "private sphere" of economic activity and the "public realm" of democracy and the state' (Picciotto 2002: 1) often obscures more fundamental questions about how we might formulate property in water as a concept that straddles both the economic sphere and the public realm in Australian cities. In urban areas, the issues of defining the public interest and state accountability, vis-à-vis the 'rights' of water users and consumers, and the environment, take on particular dimensions in the light of the regulatory changes occurring in the water supply and 'retail' sectors. Moves to deregulate urban water authorities have created models of governance that transcend the simplistic view of dichotomous public and private spheres and public/private property. Some analyses suggest that this ordering, with its terminology of the market, arises independently of the various forms of state law (Teubner 1983). Here, it is suggested that there exists a symbiotic relationship between state law and the economic regulation and techniques of the market that produces a more seamless, 'regulatory space' inhabited by both 'public' and 'private' 'actors'. Indeed, to posit a dichotomy of public and private regulatory spheres is to ignore the complexities of the interplay between private and public forms of regulation that have merged over the last few decades (see Vincent-Jones 2002: 27; and Godden 2006). On the one hand, there has been a move to allow a greater role for private actors in water-policy development and water-provision services, while concurrently governments have enhanced their auditing and monitoring functions of these private 'players' in the water sphere.

Consistent with such trends, the current emphasis on the articulation of the pubic/private modes of regulating water resources in Australia is a consequence of the national water law and policy-reform agenda initiated through the Council of Australian Governments' (CoAG) political agreement-based process and National Competition Policy (NCP). To date, these converging policy platforms have culminated in the current strategy known as the National Water Initiative (NWI) and its associated institutional and sectoral forms. Reforms of the water sector have produced a broadly-conceived, structural adjustment where the emerging governance and regulatory models share many common features with trends to privatisation, commercialisation and commodification of water in other social-democratic nation states (Bakker 2005: 544).

In particular, key components for the governance of water are a response to an increasingly prominent series of rules and principles associated with 'economic globalisation' as an ensemble of legal frameworks, policies and institutions for the promotion and protection of private property, investment and trade. Such rules and structures cumulatively attempt to fashion a global vision of economic policy, property rights, and the role of the state and government that institutionalises the political project called 'neoliberalism' (Schneiderman 2000: 84). Under the impetus of this project, an expectation arises that governments will remove regulatory restraints on the movement of capital, property, goods and services. In a corresponding trajectory, governments at all levels are under pressure to divest common resources and publicly-owned enterprises to create private property rights, and to facilitate the private supply of goods and services (Schneiderman 2000: 85). 'Neoliberalisation of nature' (see, for example, Marsden 2003) is a term coined to denote the phenomenon which is characterised by the establishment of private property rights in relation to commonly held natural resources. It is denoted by the use of the market as the mechanism for allocation of rights (or perhaps more accurately, for re-allocation of rights), and the adoption of cost and pricing measures that reflect environmental externalities. These elements are united by an overarching philosophy that environmental components are most efficiently used and allocated when regarded as economic goods. Markets, rather than being seen as contributing to environmental degradation, are regarded as a large part of the solution.

In this context, '[w]ater has been called the last frontier of privatisation around the world' (Petrova 2006: 1). Yet privatisation of water admits of various gradations which can include situations where:

1. The public entity has ownership of the resource and control over all assets and retains ultimate responsibility but allows some private minor 'service' delivery in a limited sphere such as marketing.
2. The public entity retains ownership but outsources one or more 'core' responsibilities (billing and collection, maintenance, environmental services, training, technology upgrading, procurement management, or other such tasks).
3. The public entity contracts with a private entity to fully operate, maintain, manage its water-supply system, or some significant portion of it.
4. The public entity sells the water-supply system, or a significant portion of it, to private entity (Arnold 2005: 6).

Privatisation of water resources in urban areas, therefore, could adopt any mixture of the models of resource management and outsourcing of services under a general rubric of privatisation. These configurations may or may not affect the fundamental construction of the respective 'property' in water between the state and private individuals. Indeed, Bakker (2005: 543) posits that privatisation and the introduction of markets has not been uniformly applied in converting water into a private 'transferable' commodity. In England and Wales, for example, there has been a resurgence of a mixed private/state-based regulation. Indeed, it is often overlooked that many privatisation objectives are to be achieved by applying a mixture of regulatory policy, legal 'tools' and market mechanisms that either seek to place constraints on the scope of 'state' regulation (exemplified by trends such as 'cutting red tape') or, alternatively, seek to enhance state capacity through increased private provision of services, characteristic of 'Public Private Partnerships'. Again, these regulatory models may impinge on the relative water 'property' distribution between the state and individuals, although complete transfer of public ownership of the water itself, as distinct from infrastructure, into an economic good is unusual within Australia to date (Fisher 2004: 201). Surprisingly, even in the United States, the majority of water resources remain in public control, although various forms of private rights are recognised (Glennon 2005: 10). Nonetheless, the end result in recent times has been to perceive the function of the state in western democracies in an increasingly constrained manner. Accordingly, there has been pressure to progressively narrow the powers of the state with respect to the control and management of water; firstly in response to development-oriented resource-based approaches, and more latterly under globalisation and privatisation trends.

However, as Fisher (2007: 122) notes, current water-reform strategies under the NWI employ a range of legal, regulatory, market and behavioural models to achieve the strategic outcomes, all of which will have differing degrees of legal enforceability. Thus, it is difficult to identify a discrete boundary between the public, the private and respective property 'rights' without reference to specific institutional and governance forms, notwithstanding the claims that are made about the pervasive privatisation of water.

Increasingly, nonetheless, in the utility and resource sectors, there are emerging forms of water regulation typified by groups of public and private actors who are linked through an 'identified' policy problem or objective, such as water security. By virtue of the manner in which the 'problem' is defined, it is seen to necessitate joint participation and a joint expenditure of resources. Typically, such inter-organisational structures are expressed in reinvented forms of agreement, which are then given legal expression in the form of contracts, partnerships or even as joint ventures. Innovative, institutional and legal governance forms have appeared in Australia with blended corporate structures where private law models of the corporation are adapted to public functions in structuring the functions of bodies such as water authorities (for a discussion of these trends within the European Union, see Teubner 1983). As noted, where both approaches converge is around a construct of market environmentalism — 'a mode of resource regulation that promises both economic and environmental ends via market means' (Anderson and Leal 2001).

However, while adopting joint objectives, market environmentalism is not uniform in its distributive consequences. The conjoining of economic-efficiency outcomes with environmental goals under market environmentalism typically occurs at a highly aggregated level, typically expressed as consumer preferences or 'highest and best use', which may ignore local and spatially skewed impacts. Further, the apportioning of risks and returns on joint arrangements and expenditure under the processes of risk-spreading that operates within the financial and security sectors, still suggests uneven responsibilities for the public and private 'actors' in the water sector. These effects can be seen in both environmental impacts and in financial obligations — an important consideration given the large capital costs associated with water-infrastructure development. Any potential distributive inequalities are exacerbated as financial arrangements, despite some attention to triple-bottom-line sustainability objectives, remain more traditionally structured around 'purely' economic objectives. As the head of project finance *nabCapital* commented recently:

> [F]rom a banking perspective, there's a tremendous opportunity [in construction of water infrastructure], particularly if it is financed in a PPP-style arrangement. The capital markets, the banking markets and equity markets have a strong appetite for this type of risk. They view

it as essentially government off-take risk, with long stable cash flow. (See Kellerman 2007: 24)

Whether it is appropriate for governments faced with severe water-supply situations in urban areas, and in the light of long-term drought exacerbated by climate change, to seek to mitigate the risks of water supply by various 'privatisation' or market strategies goes to the heart of debates about efficiency and sustainability of water regulation in urban areas. These trends also have significant implications for rural areas as many of the 'water resources' that governments offer as the basis for water-supply projects will be physically located in rural regions. Most importantly, many of the repercussions of these projects will be experienced in the environment and communities of those rural areas.

Arguably, in the context of recent government policy on water security in urban areas, we need to invert the well-worn adage of Adam Smith in the *Wealth of Nations* to examine not just the invisible hand of the market but the 'invisible' hand of government in the market. Law, whether legislation, or through court-based processes of dispute resolution, is a key mechanism for governments retaining a hand in the market, even if this presence remains largely unacknowledged. Thus, in concert with such a regulatory approach, the role of law is transformed from its traditional command-and-sanction position to be held in tension with other policy drivers such as 'market forces'. Accordingly, within this paper, the theoretical context combines an approach which posits law (and regulation) as an expression of normative 'communications' between social, economic and political groups or networks, and which recognises that law provides only one, albeit a very significant, technique that is deployed either explicitly or implicitly to achieve those normative objectives (for a discussion see Black 2002: 5–6). In a former era we might have referred to the development of 'policy' and then its implementation. Yet law sits awkwardly across such a division. Law is a major influence in the creation of the governing normative order but it is also key to its implementation. Law is central to the institutional design that structures discourse around water (see Dryzek 1997: 109) and also a major instrument to effect the implementation of those normative objectives. If that sense of law is accepted then it is integral to a more complex and reflexive sense of water 'regulation' that has come to the fore within Australia. In turn, this impinges upon how property as a legal construct needs to be understood, both as a normative expression and as a 'tool' of implementation for policy objectives.

Water as the 'property' of the state

In any such examination it is important to acknowledge that even the most expansive models of the commodification of water and risk 'sharing' still assume a situation of initial state control or 'ownership' over water in its various forms. Property constructs are variable and, as Fisher (2004) notes, can range from property held by the state, an individual, a community or an 'unowned' commons. Whether the posited 'unowned commons' of economic-resource theory can exist in reality is a moot point (for a discussion, see Fisher 2004: 201). Here in Australia, as noted, most debates about the degree and nature of privatisation in relation to water coalesce around the character of 'public ownership' rather than querying the initial vesting of the resource in the state. Typically in natural resources contexts, such vesting occurs through statute. In this regard, the Australian High Court has been called upon to determine the nature of Crown [Government] 'ownership' of natural resources in the seminal case, *Yanner v Eaton*. In that case, the Court accepted that:

> 'Property' does not refer to a thing; it is a description of a legal relationship with a thing. It refers to a degree of power that is recognised in law as power permissibly exercised over the thing. The concept of 'property' may be elusive. ... Much of our false thinking about property stems from the residual perception that 'property' is itself a thing or resource rather than a legally endorsed concentration of power over things and resources.' (*Yanner v Eaton* at paras 17–18 per Gleeson CJ, Gaudron, Kirby and Hayne JJ).

While recognising the integral link between property and its legal designation and enforceability, the High Court clearly distinguished Crown ownership or property in a resource as being distinct from 'private property'. The Court concluded that the Crown's property was an atypical form of property, (*Yanner v Eaton* at para. 26 per Gleeson CJ, Gaudron, Kirby and Hayne JJ). Property in the Crown is, in effect, the mechanism by which the state asserts regulatory control over access to the resource. Accordingly, the powers of the Crown must be interpreted in the light of the legislative scheme, which vests that property. While *Yanner v Eaton* was concerned with the 'vesting' of wild animals in the control of the State under Queensland statute, the judicial reasoning has particular consequences for current debates surrounding privatisation of water. The conception of property as pertaining to a socially approved 'power' over resources, and the distinction between state 'regulation' and private rights (otherwise known in formal legal terms as 'the beneficial interest'), highlights the public-interest dimensions of Crown property or 'ownership' of water resources. Thus, this 'socially endorsed power' could comprehend a range of public-interest criteria including conceivably 'a duty' to ensure minimal rights to water, environmental protection and intergenerational equity (Fisher 2004: 209).

In Victoria, for example, there has been an attempt to clarify the respective interactions of state 'control' of the water resource and arrangements with the private sector through a constitutional entrenchment of the public responsibilities for water (Sections 96, 97 *Constitution Act* 1975 Vic). Such a view of property, state control and private property that emerges from the foregoing analysis represents a more nuanced approach to 'property rights' and the gradations of 'privatisation'. Many analyses of rights in water tend to concentrate upon the imposition of private property rights without proper regard to the wider context in which such rights exist. These analyses typically move to focus upon 'private property', security, trade, transfer and transactional arrangements without duly recognising the relationship of such rights in the context of state 'ownership'; which, in turn, must be conceived more widely than simply as a conduit from the vesting of the water resource to points of allocation to individuals. What then is normative role of the state in articulating property in water? For whom does the state 'act' and in what capacity? Many answers are conceivable: Current urban water consumers? Current business interests? Water users beyond the immediate urban spatial scope? Future generations of water users? The environment?

Consequently, if the state is held to hold an atypical 'common property' in water, it becomes important to explore how water might need to be considered under a broader rubric of state responsibility as part of its 'invisible' hand in the market. Clearly there are public-interest outcomes that markets, when constructed as the aggregation of consumer preferences, simply cannot achieve. Typically, markets do well in allocation and efficiency but less well on issues of broader social and spatial distribution and equity (Bakker 2000; for a contrary view, see Heaney *et al*. 2006). One of the central arguments made for adopting a market-based or property-rights approach to water-resource regulation is that markets allocate scarce resources more efficiently than the public sector. Public-sector limits on scarce resources, such as water restrictions imposed under legislation, it is argued, are inefficient as they impose the same level of restrictions on all water users without regard to the 'value' that users may have for the water. Some arguments are made that these government legislative measures are appropriate only as interim 'emergency' instruments (Quentin and Ward 2007: 8). However, it is suggested that a market-based approach allows those who have a higher-use value for water (and are able to pay to higher prices) to use scarce resources more efficiently. Property rights, it is suggested, provide an incentive to use the water resource more efficiently as other users can be excluded (Posner 1998). While this is an over-simplification of complex arguments, one major assumption not addressed is the assumption of scarcity as a 'natural' phenomenon. By contrast, it is suggested here that scarcity and use value are notoriously relative constructs. Indeed, it might be argued that private property, rather than being the 'solution' to natural scarcity, simply puts in place a private law regime for rationing (that is, state supports private law regulatory forms to ration the resource). Whether the rationing takes place according to a public-law or private-law regime, necessarily law is involved, rather than the situation being one with a pure market-based 'solution' and the other being a 'state' law-based regime. The articulation of the normative objectives for these legal regimes then becomes critical in any determination of the relative merits of the regimes for ensuring long-term water security and sustainability.

In particular, it is argued that privatisation may produce a 'win-win' between the parties to any water value-exchange transaction but fails to deal with third-party effects, including environmental degradation (Glennon 2005). Moreover, even if we concede that the role of governments should only be to intervene in situations of demonstrated market failure, the range of situations is potentially quite wide. Accordingly, the equation of efficiency in water-resource allocation with Pareto optimality (that is, allocating water to its highest value use) is sufficient to satisfy social-welfare objectives needs further exploration. In this context, the view that water could be conceptualised within a rights-based paradigm may offer a useful counterpoint.

Water as a human right or social need?

At an international level there is an emerging dynamic around the right to water (Hildering 2006). Arguments have been raised that access to a basic water requirement is a fundamental human right and this is implicitly supported by international law, declarations, and State practice (Gleick 1999: 2). Various formulations have been articulated such as that of *WaterAid*, which suggest that: 'The right to water is the entitlement of everyone to have access to sufficient, affordable, accessible and safe water supplies and sanitation services. It places an obligation on states progressively to realise the right to water for all people without discrimination...' (WaterAid 2007). To date, though, there is no clearly agreed upon and enforceable right to water per se.

At its most expansive, the right to water could capture aspects of human rights based around fundamental 'needs' as well as the protection of ecosystems. A rights-based approach could also subsume an ecosystem approach as:

> The term 'right to water' does not only refer to the rights of people but also to the needs of the environment with regard to river basins, lakes, etc. Realistically, a right to water cannot be secured without attention to this broader context. A failure to recognise water as an environmental resource may jeopardise the rights-based approach, which views water primarily as a social resource. (Scanlon *et al.* 2004: 22)

Much of the impetus for recognising water as a human right, to ensure a guaranteed level of access to clean, safe water supply, developed from the view that if water is regarded as an economic good then it may lead to a denial of access where some peoples cannot afford the increasing charges associated with full economic recovery of water-provision costs (Bluemel 2004: 1). Further, if such a right was accepted, then its enforcement would presumably oblige the State (and potentially other parties) to provide water when minimal access is lacking (Miller 2005: 8).

While the momentum generated at an international level to recognise rights to water, either as a stand-alone right or as a complex of human rights and environmental protection, has been significant in shifting attitudinal paradigms about the value of water, the lack of overt enforceability of such 'rights' remains a barrier to practical effectiveness. Explicit constitutional recognition of rights to water, as it occurs in countries such as South Africa, clearly enhances the legal status of such rights and provides a more discrete mechanism for enforcement. Even so, there remain substantial limitations upon directly enforcing any wide-ranging right to water, not least being the high costs of pursuing enforcement through any court or tribunal system. As Fisher (2007: 16) notes, a right or, perhaps at best, an expectation to receive water is conferred mostly in relation to the supply of water for direct consumptive uses in a residential setting in urban areas. Typically, also, any such right is qualified as it will need to be interpreted in the light of corresponding duties cast upon the suppliers of water which are set within a wider legislative and strategic framework that encompasses water policy, strategies, discrete water plans and an integrated array of water rights and obligations. In each instance, the rights and duties are unlikely to be absolute but may be couched as overriding statutory objectives (Fisher 2007: 122). If private entities are held to have such duties in situations where water resources or some essential component of the delivery infrastructure is held privately, the potential for difficulties of implementation of any 'right to water' are exacerbated. Therefore, to date, the main thrust of thinking about the responsibilities that are generated in relation to water has focused on the public dimensions (Fisher 2007).

Clearly also, the formulation of a right to water gains increasing urgency in conditions of scarcity, exacerbated by climate change, where even the most basic access to drinking water may be under threat. However, whether a guarantee to water should be designated as the driver of policies such as 'water security' in Australian urban areas is much more contentious, especially where any rights beyond direct human consumption and basic hygiene are being formulated. In Australia, most basic 'rights' to water are currently articulated in legislation, including the adoption of specific legislative safeguards to ensure access to water for vulnerable domestic water 'consumers'. For example, in Victoria the Essential Services Commission, established under the *Essential Services Act* 2001, regulates water prices. Included in the overarching objectives of the governing legislation is a clause specifying that one goal of the Commission is 'to ensure that users and consumers (including low-income or vulnerable customers) benefit from the gains from competition and efficiency'. With many basic 'rights' to water already largely in place, debates about 'rights to water' within Australia in future scenarios of water security are more likely to be understood in terms of more discretionary water uses. These might be argued on the basis of social or cultural needs, as exclusions from an increasingly stringent 'user-pays', cost-recovery system of water pricing being introduced through market mechanisms. Typically these needs-based arguments are fairly well accepted in Australia on the basis of consumer- protection or social-welfare outcomes, rather than a more specific human rights categorisation.

The environmental protection component of any putative human right to water is more difficult to prescribe, particularly for urban water. Typically there will be vast distances between the ecosystem and its in-situ functioning and the provision of water to far-off urban water consumers. Arguably the preservation of ecosystem functioning should be a first-order 'right' protected prior to any consumptive use of water. However, despite adoption of sustainability objectives across many areas of water management policy and legislative reform within Australia, environmental water is rarely accorded this priority (Foerster 2007). Typically, under current NWI reforms, environmental water is regarded as requiring the same degree of 'certainty' and legal status as consumptive entitlements; a measure that is welcomed. Yet, in practice, particularly in severely over-allocated river systems, this has still meant that in many instances the provision of water to support ecosystem functioning occurs after allocation to consumptive users as existing entitlements. Environmental water releases, generally rely on highly discretionary decision-making by relevant government ministers with few direct accountability provisions. While market environmentalism would suggest that both efficiency and environmental objectives could be simultaneously achieved, drought and climate change have highlighted the practical impossibility of resolving the competing priorities for water in many catchments across south-eastern Australia. Only if time and spatial scales for calculating efficiency are abstracted and aggregated as a general overall value, apart from the local and contingent uses of water, are such objectives likely to be synchronised. On the other hand, individual efficiency and utility maximisation for water use may be offset against dispersed community amenity loss. (For the contrasting view that there are greater welfare costs associated with water restrictions, see Quentin and Wood 2007.) In other words, local environmental degradation in a given area may be 'offset' against efficiency or security gains across the catchment or indeed across a rural-urban region. Typically the water security 'offset' has been in urban areas.

To date, the competing priorities of consumptive water use and environmental water have been primarily seen as rural issues. Yet if we take a more holistic and spatially extensive view of the ecosystems that 'deliver' water to urban residents, then clearly urban water consumers compete for priority under assignments of water 'rights' that cover a much wider spectrum. If governments increasingly seek to provide water security for urban areas, and if 'barriers to trade' are to be progressively removed to facilitate rural to urban water trade, then urban water consumers, whose water supplies are insulated in particular ways, may form a significant competitor for rural water. Moreover, whether a rights-based or social-need argument for urban populations might provide a rationalisation for urban water supplies to be guaranteed vis-à-vis environmental functioning or rural uses of water is an issue to be explored further as part of an examination of the temporal and spatial patterns of market environmentalism as it impinges upon current policies for urban water management.

The Australian urban context

The processes of water regulation that have emerged in Australia, while influenced by international trends across the spectrum from market environmentalism to human rights discourses, exhibit some distinctively Australian characteristics. These patterns reflect the historical and contingent importance of water as a 'resource' within Australian society and the economy. In the first phases of the CoAG and NCP national water-reform agenda, urban water issues were clearly eclipsed by concerns over the seemingly intractable decline of rural water resources. Water reforms were first sparked several decades ago by heightened concerns over salinity, together with a growing acceptance of the long-term adjustments occurring across rural industries, which traditionally relied on extensive water use. Severe water shortages over the last few years in traditional urban water supply in south-eastern capital cities and in Perth, have refocused government and community attention on water in, and for, cities. Governments, which in the past had enthusiastically embraced a hydraulic model of water supply (Bakker 2005) largely predicated upon high levels of technological infrastructure development, were seemingly uncomfortable with stringent demand-side measures, such as staged urban water restrictions. While such demand-related measures were instituted (see, for example, D part 3 division 1A *Water Industry Act* 1994 Vic), governments across all states seem to find long-term restrictions on water use for urban communities unpalatable. In a reversion to earlier dependencies on water-resource development models characterised by high levels of technological complexity, many governments have instituted infrastructure-based supply-side 'solutions' under the rubric of ensuring water security for urban areas.

In this context, arguably, what has occurred is that despite a purported reliance on market mechanisms to institute behavioural and structural change in water use, the neo-liberal reform agenda has been amalgamated with earlier models of state-based infrastructure development to form a hybridised model of state control, privatisation and market forces. While the potential for transactional-based market 'efficiency' approaches still exists, many recent policy choices privilege water-supply infrastructure development, albeit in concert with market mechanisms such as water trading. New species of water 'rights' are being advocated, such as the right of access by prospective businesses to public water-infrastructure development either as 'third-party rights of access' to existing infrastructure or through private 'participation' in new projects for water-infrastructure development. Such approaches have direct ramifications for the balance between state 'ownership' and control, and private 'property' in urban water. As Gray and Gardner (this volume, Chapter 7) note in regard to urban sewer mining and wastewater treatment: 'The recycling of human wastewater is being reinvented as both an environmental and commercial opportunity that can be facilitated by giving "third-party" access to established public-sector infrastructure and — importantly — to the sewage!!'

Third-party access issues and the associated implications for conceiving property in urban wastewater are dealt with extensively in Chapter 7. Therefore this chapter does not canvass those points. Rather, the chapter addresses the manner in which recent policy decisions and emerging legal frameworks for urban water reinscribe but also transform earlier legal patterns of water-supply resource development within Australia. Models of state resource development and access have been modified, but not displaced, by market mechanisms and trends to privatisation. Water regulation exhibits an accretion of various layers of regulation and law rather than a situation where one model of law and regulation is sequentially replaced by another; notwithstanding the extensive law-reform processes that have been initiated in recent decades. To understand this accretion process it is necessary to understand how rural and urban areas were integrated, but separately regulated, under the earlier resource-oriented modes of water law and policy. Arguably, current models of water-resource development founded on technological infrastructure reinstitute interdependency between rural and urban — a factor which remains only partially acknowledged in emerging policy and legal frameworks.

Within Australia, historical patterns of water use, post colonisation, were linked to prevailing land-use settlement patterns. Urban development was strongly influenced by a colonial, and then national, agenda of primary production, and an export economy funnelled through the major port and infrastructure nodes represented now by the major metropolitan capital cities. The patterns of production introduced under the colonial enterprise whereby Australia, together with many other parts of the colonised world, supplied food and fibre to a rapidly industrialising Europe were premised upon the 'colonial earth' and its resources largely being treated as 'free goods'. This was accompanied by economies of scale in the infrastructure development under the increasingly sophisticated regime of land and resource administration and utilisation (Godden 1997).

Integral to such an emerging administration system and resource-exploitation program that produced the 'classic' nineteenth-century colonial model of centre-periphery governance focused upon the colonial capital cities was the necessity to dispense with the riparian doctrine of water rights. Key aspects of the doctrine initially centred upon allowing 'use' of water by upstream users so long as there was not undue interference with the 'economic utilisation of water' by downstream users. At its very basis, it is an appropriation model that adopts a very simplistic instigation of individual rights and associated allocation principles over a common resource.

Within Australia from the late nineteenth century onward, there was a progressive move to replace riparian doctrines with a statutory framework for water regulation, predominately to promote inland settlement based around the implementation of irrigation schemes, largely but not exclusively in the Murray-Darling system (Clark 2003). In urban areas, too, the force and influence of bureaucracies grew and a corresponding statutory and institutional basis for water-supply provision was progressively implemented (Powell 1989; Davison: this volume Chapter 3). Trends to institute statutory forms of water governance and associated public utilities required an articulation of the legal foundation for such governmental control in the statutory vesting of water resources, which occurred in concert with a move from smaller, private sources of water supply.

Urban models of water governance

In turning more directly to the urban sphere, much of the impetus for the developing statutory models of water governance lay in public-health concerns over water supply. Further, as Davison contends in this volume, the identification of cleanliness with 'flushing away' wastes was important for the expansion of extensive sanitary-waste systems that were instigated along British models. In concert with the genesis of much early town planning law, the development of urban water supply and sewerage systems and their associated institutions in Australian cities gave a particular spatial form to these cities. Increasingly water supply was being drawn from 'clean' areas beyond the cities. For example, Melbourne has an extensive area of 'closed catchments' for water supply instigated in the late nineteenth century and early twentieth century that reflected this impetus (Powell 1989). Similarly, wastes from cities were to be carried well beyond what, at the time, were the perimeters of the cities (Davison, Chapter 3). The influence of water utilities in many cities was such that urban form was shaped by the intricacies of the distribution systems of water supply and sanitation infrastructure. Large statutory water authorities became virtually autonomous entities, in many instances having their own unique enabling legislation and institutional regimes. The Melbourne and Metropolitan Board of Works (MMBW), for example, wielded enormous power in political and economic spheres. At one stage it assumed responsibility for the statutory land-use planning process operating within the metropolitan area and set the parameters for the physical and social development of the city well into the late twentieth century

(Powell 1989). The prominence of the water authorities, both urban and rural, reflected the identification of national goals with a virtually unconstrained exploitation of water resources throughout most of the twentieth century. In the post-war period, infrastructure, including water supply and sanitation delivery, came to be regarded as a public good requiring monopoly provision by the state (Connell 2007). Water shaped city form as it still does in a myriad ways in the twenty-first century (Syme, this volume Chapter 6). Most importantly also, it shaped the interaction of capital cities with rural areas by continuing a trend whereby 'resources', including water, were drawn from the rural hinterland, and externalities, such as wastes, were visited upon rural areas. However, this neat systemic view represents an oversimplification, as there were many points of resistance to such 'transfers' and political compromise and compact reached for exemptions from these dominant trajectories. Yet this underlying pattern of rural–urban interaction has not been directly displaced.

Current urban water reforms

Urban water-law reform, in concert with broader trends, is premised upon a move away from traditional statutory-authority models of water provision and regulation. Primarily, the changes constitute a move from monolithic public-authority models towards the creation of semi-autonomous water 'production' and distribution entities and 'separate' regulatory institutions. These entities are to function in a market economy; albeit in which the state retains significant control over the functioning of the market. In some instances, the state may constitute itself as an indirect 'player' in the market through shareholding in water utilities. Increasingly also, in the urban context, the normative order regulating water increasingly incorporates not only formal laws but a plethora of less-formal standard-setting, behavioural-change models and the development of monitoring and compliance processes, all administered and enforced by a mixture of public authorities, independent agencies such as price regulators, and private 'consumers.' (Parker *et al.* 2004: 1). Regulators operate increasingly in a pluralistic setting where certain state functions are shared with, or devolved to, private interests (Keohane *et al.* 1998: 314). Thus, while market-based economic 'theory' and its assumption of deregulation have been highly influential as an impetus for recent water-law reform in Australia, regulation of urban water presents a more complex picture. Externally-defined purposes set by agencies such as the National Water Commission (which administers the NWI) are incorporated into the structural shift from predominantly statutory-based water authorities to a situation of water 'suppliers', 'consumers' and, potentially, an expansive market system of water trading.

Victorian metropolitan water reforms

Victoria provides one such example. That state embarked upon a restructure of the metropolitan water industry in the 1990s that aimed to 'introduce commercial measures [to] improve customer services…' (Office of State Owned Enterprises 1995: 1). The main component of these commercial measures was division of the MMBW, the former all-encompassing metropolitan statutory water authority. The authority was separated into the corporate entity — Melbourne Water (see *Melbourne Water Act* 1992), whose functions largely became the provision of bulk water and wastewater disposal — and three state-owned enterprises that assumed responsibility for the retail supply of water (see *State Owned Enterprises Act* 1992). The state-owned enterprises were established in a corporate form, 'as this best replicates a commercial operating environment.' (Office of State Owned Enterprises 1995: 3). The commercial competition, however, was a 'competition by comparison and benchmarking' as the metropolitan region is divided into geographically discrete areas in which each water enterprise is the sole provider. Each water business functions under a separate operating licence. This is in contrast to the earlier system where all supply and distribution functions were consolidated in a single statutory entity. Water pricing in relation to the water retailers is set by an independent regulator, the Essential Services Commission. Detailed requirements for the economic regulation of the water industry are provided under a Water Regulatory Order. This model for the metropolitan water industry shares similar features with regulatory models adopted pursuant to competition-policy objectives in utility sectors across Australia and overseas.

The Victorian government, in line with water-policy initiatives at a federal level, released a White Paper, *Our Water Our Future*, announcing significant water-law reforms in mid 2004. Amendments to the *Water Act* 1989 (Vic) and the *Water Industry Act* 1994 have progressively introduced a raft of CoAG and NWI reforms which built on the extensive legislative reforms already undertaken in metropolitan water supply. While much of the reform agenda is directed specifically to rural areas (for example, 'unbundling of water entitlements'), there are ramifications for urban areas. In addition, further changes have been introduced into the general regulatory structure with the move to instigate water authorities as corporate entities under the Corporations Law. As of 1 July 2007, all water authorities in Victoria were restructured and are now classified as corporations. This change arises from the *Water (Governance) Act* 2006, which came into operation on 18 October 2006. Of significance is that, as corporate entities under the Corporations Law, there are enhanced reporting and accountability requirements for these entities. Enhanced accountability is required to shareholders — which in this instance is the state government. Exactly how this blend of private corporate responsibilities may sit with public duties of water 'ownership', which to date remain vested in state control, remains to be seen. Such merging of public/private spheres is seen as evolving in the context of a wider move to open a space between 'the old dichotomies of state, market, public, private, local, global' (Considine 2005: 1).

Reform of the Victorian water sector, though, in some respects has not reached the level of privatisation of water and water infrastructure obtained in some other jurisdictions. Further, until recently, it was clear that there were significant political incentives to retain substantial governmental ownership and control of the water-resource sector given constitutional entrenchment of public provision of water services (ss 96, 97 *Constitution Act* 1975 Vic). However, the Victorian Competition and Efficiency Commission, which was established in 2004 with a brief to 'improve the awareness of, and compliance with, competitive neutrality', has undertaken an 'inquiry into the reform of the metropolitan retail water sector'.

The inquiry made recommendations regarding:

- the best structure to allow for the efficient and least-cost provision of Melbourne's water-supply upgrades, as well as ongoing safe, reliable and sustainable water and sewerage services to Melbourne;
- options to reduce costs of the metropolitan sector whilst maintaining and improving the level of service over time and ensuring it remains innovative and financially viable;
- the broad staging and timing of any proposed structural reforms to the metropolitan water sector; and
- any related improvements to governance and industry structure in the context of the Government's Water Plan and climate change.

(Victorian Competition and Efficiency Commission 2007: 6)

The context for such an inquiry is 'The Next Stage of the Government's Water Plan' (Department of Sustainability and Environment 2007). A principal component of the plan was the announcement of major water-supply projects for Melbourne. These infrastructure projects include reconnecting the Tarago Reservoir (15GL), the Sugarloaf Interconnector (pipeline) (75GL) and Australia's largest desalination plant (150GL). The projects are accompanied by large projected increases in average consumer water costs in Melbourne by 2012. Metropolitan water authorities — Melbourne Water, as the wholesale supplier, and three retail water companies, City West Water, South East Water and Yarra Valley Water — have finalised draft water-pricing proposals involving price increases of between 100 per cent and 140 per cent (Victorian Competition and Efficiency Commission 2007: 7).

At first instance, these reforms seem to be concerned principally with efficiency parameters and the operation of market mechanisms, such as water-pricing measures, within the urban water-industry sector. But on closer inspection it is apparent that the issues of water pricing and economic performance of metropolitan water retailers are being assessed in light of major capital investment, in water infrastructure projects for water supply, located predominately in non-urban locations. Moreover, in light of recommendations to identify objectives such as 'the best structure to allow for the efficient and least-cost provision of Melbourne's water supply upgrades', are these structures to be considered as 'any related improvements to governance and industry structure in the context of the Government's Water Plan and climate change'? If so, at what scale and across which spatial and temporal dimensions will such objectives operate? Are the urban–rural interactions to be considered? The potential scope of any inquiry into metropolitan water governance reinforces the view that urban water regulation cannot be regarded as operating in isolation from surrounding regions. More controversially, perhaps urban water pricing should not be set without regard to the externalities and third-party effects that will occur in those regions, including potential widespread environmental effects. To understand potential ramifications, it is necessary to consider intra-urban water management and then regulation of water at the interface of the urban and rural.

Securing urban water supply

Intra-urban management

As Syme notes in this volume, urban water utilities have been conservative in their reaction to changed demand-and-supply options. The focus is on technological solutions, and the identification of new sources of water supply with well-established approaches to demand management. Some attention is given to water-sensitive urban design and the incorporation of externalities into pricing and cost-benefit analyses, but there are large gaps in achieving sustainable management. While sustainability agendas require more innovative adaptations to changing water availability, to date, the issues of increased scarcity in urban water have been largely managed on an intra-urban basis by regulating through the imposition of water restrictions and efficiency incentives, based largely on pricing measures. Some commentators point to the success of demand-side measures, but clearly the impetus is shifting yet again to give priority to supply-side 'solutions' for Australia's cities. If supply side-options are in the ascendancy, the issues of property, access to ware resources and rights to water again become of central importance.

Debates about the utility of property concepts and market mechanisms in achieving the goals of sustainable water use and efficiency have, to date, largely focused upon the rural sector. Yet many of the questions that arise about the balance between private rights and the public interest have similar resonances in an urban setting. Water-scarcity issues have impacted (impact) all our major cities. Typically, while demand-side measures such as water restrictions have been implemented, and there is much exhortation to change ingrained social practices such as showering, the 'business as usual' scenarios have prevailed. The hope of technological fixes of new dams, desalination plants, water recycling and innovative technologies are being held out as solutions to expand options for metropolitan water supply.

Recycled water, at first instance, appeared to be the panacea to the problem of growing urban water scarcity. While much potential obviously exists in wastewater recovery and treatment, considerable cultural resistance has emerged to recycled water for residential purposes (Hurlimann 2007). Further, water-quality standards require intensive water treatment, which means that in many instances it is not cost-effective. Recycled water systems, where instituted, often have received public subsidy; although one might query whether the full costs of environmental externalities, including energy demands and greenhouse gas emissions, have been costed into other alternatives vis-à-vis recycled-water pricing. Nonetheless, attention is being directed now to supply infrastructure development that can look to 'fresh' sources of water. In concert, there exists the potential to use water 'property' trade/exchange models more directly.

Scarce water supply has been addressed in recent policy paradigms — particularly in the rural sector — by use of allocation mechanisms; that is, to institute entitlement-based regimes and potential re-allocation of water to the so-called highest and best use through the property/contractual/trading process. However, the highest and best use of water, if taken to its 'logical' conclusion can operate across a variety of spatial scales. Rural–urban water exchanges clearly fall within the possible scales.

Rural-to-urban water trade models

On an ad hoc and limited basis, it is clearly possible for urban dwellers to purchase water from outside urban areas. Trade in rural water to supply urban swimming pools (*The Age* October 2007: 1) is just one example. Arguably, such trade meets water-efficiency objectives and it emulates models borrowed from the United States. As Glennon (2005: 1902) notes, '… the best way to reform agricultural water use in the United States is to give farmers a financial incentive to use less: let them sell water to the cities.' Indeed the bulk of water trade in the southwest corner of the US is rural-to-urban water sales. Such potential clearly exists in Australia to facilitate an increased rural-to-urban water trade based on private modes of exchange through contractual and property regimes. Young (2007: 86) argues that even with the projected five million increase in the Australian population over the next 25 years — which will occur mainly in urban areas — any estimated 'transfer' of water from rural to urban areas under such a process will amount to less than 1 per cent of the amount of water that is extracted for use. Such water trade would not be 'unidirectional', with water flowing to cities and taxpayer and water consumer 'investment' flowing back to regional areas. He cites existing urban subsidy of rural environmental rehabilitation such as the natural heritage trust funds as a model to be emulated. Moreover, Young argues that such rural-urban water transfers under a willing seller/willing buyer formula would provide a net benefit for investment in water even while recognising the need to consider environmental externalities and third-party effects (Young 2007: 90).

However, the costing and assumptions for calculation for such environmental externalities are rarely made explicit (see, for example, the multi-criteria analysis employed to determine options regarding the Goulburn Campaspe Link GHD 2006). Moreover, perhaps the most significant assumption of the calculation is that such benefits would accrue on the basis that all the issues at the heart of the NWI process would be addressed, such as the return of all surface and groundwater in rural areas to a healthy ecosystem state. Clearly to date, there has been significant achievement on the water-use efficiency measures and the instigation of processes to support water trading. By contrast, there has been very limited progress on the actual return of waters to over-allocated rivers and water systems in regional Australia. Even where environmental water has been allocated, there are substantial impediments to its effective implementation to support ecosystem functions (Foerster 2006; Ladson and Finlayson 2002). The view that equivalent 'benefits' could be achieved by using water from rural areas to sustain urban parks on the basis of satisfying consumer preferences for 'greenery' or 'vegetation conservation' or for urban wealth generation and employment gains would suggest a fundamental problem with a market-environmentalist approach. Water, if we accept its fundamental ecosystem qualities, is not an infinitely substitutable and transferable resource. It has localised environmental, equity and distributive-justice implications.

Arguably if we accept the exchange value or 'substitution' involved, the optimal approach might be to dispense with living urban parks and gardens and move to plastic trees. Clearly water, in situ, provides many important social cultural and economic values that cannot be captured by property and exchange constructs as these are currently conceived under market environmentalism. Barriers to trade can alternatively be called legitimate policy and legal constraints in the public interest, which satisfy various public duties entailed in state 'powers over water'. Moreover, arguments that water 'holding' should be free of all restraints and barriers to trade simply does not accord with the substantial constraints and mixed regulatory objectives that operate with land holding as a form of property right. Land holding and water rights have been progressively disaggregated, perhaps most strikingly in urban areas where water is supplied as a 'service'. Yet water remains embedded in its social and physical context and, like land, must fulfil a range of, at times, competing objectives.

Further, arguments have been raised that property regimes should be excluded in situations of risk where there is the possibility of irreversible harm. Arguably, under the impact of climate change, many aquatic ecological systems and other water-sensitive biodiversity will approach the point of irreversible harm quite quickly, especially if water is drawn from already stressed rural areas to urban areas in order to secure water supply. Risk-spreading operates in a highly skewed manner, with public 'actors', like the environment, bearing inordinate proportions of the risk of climate change, which ultimately will become an intergenerational debt. Similarly, attention needs to be directed to the social and cultural impacts of rural–urban water 'transfers' beyond the efficiency incentives that potential sales of rural water to urban areas may comprehend. For example, one scenario to analyse the efficiency of urban–rural water trading states: '[T]he technical feasibility and environmental implications of connecting each of the above cities with their region and with neighbouring regions is not assessed' (Young, Proctor and Qureshi 2006: viii). Such exclusions from assessments of shadow pricing for urban water would seem to leave out major cultural costs and externalities that may be integral to determining the long-term value of water.

Institutional forms of rural-to-urban water exchange

To date, any major scheme to institute rural-to-urban water trade remains small scale and within the scope of significant controls on movement of water out of rural catchments. Indeed, in Victoria there remain limitations on the extent of water 'property' that can be held independently of land holding, despite moves to separate land and water entitlements and to 'unbundle' water holding into water shares, water-use licences and delivery charges. The issues of speculative accumulation of water rights in any development of secondary markets in water, together with potential for monopoly controls and price setting, are recognised as legitimate reasons to provide limitations on freestanding water 'rights'. Clearly the hand of government regulation is in the market to achieve particular public-policy outcomes.

Significantly, therefore, it appears that the trends to institute market environmentalism sit alongside much more mixed regulatory objectives. What appears to be occurring in the water-security context is an amalgamation of the resource development/infrastructure mode of water regulation, with 'public', rather than market, mechanisms of exchange. To date, the exchange is not the typical inter-party private mode of commercial water trade but one instituted by the public sector, albeit with considerable private participation in infrastructure development. Under the rubric of securing water supply for urban areas, governments again have adopted a classic resource model whereby regional resources are drawn to the centre from the periphery. This mode is evident in the latest stages of the Victorian Government Water Plan with its focus on three major infrastructure developments, all dependent upon drawing water resources from regional areas. The food-bowl modernisation project and the associated 'pipeline' exemplify this process.

Food Bowl Modernisation Project

The Food Bowl Modernisation Project (FBMP) forms part of a number of major infrastructure-development projects intended to augment water supply for Melbourne and other regional centres. The FBMP is intended to provide up to 450 billion litres of water at a cost of $2 billion. The first stage will cost $1 billion and projected water savings of up to 225 billion litres are to be shared between Melbourne, irrigators and the environment.

The focus of this project is a substantial upgrade and redevelopment of irrigation distribution systems in the Goulburn Valley. This will include channel automation, piping, channel linking and metering of the Goulburn Murray irrigation systems. The second major component of the project is the construction of a pipeline that will allow delivery of Melbourne's share of the water savings to be brought to the city. The State Government will provide $600 million towards the project. Two water authorities, Melbourne Water and Goulburn-Murray Water, will provide $300 million and $100 million, respectively. These funding arrangements for the project reflect both the public interest in improvement of outdated and inefficient water-delivery systems (efficiency improvements from 70–85 per cent are projected to be achieved) and the perceived need to augment Melbourne's water supply. Indeed, to that end, it is proposed that Melbourne will receive its 75 billion litre share of water savings in 2010, ahead of the both irrigators and the environment.

This proposal is controversial on a number of fronts. Farmers in the region are opposed in principle to the movement of the region's water to the city. In addition, if savings are not fully achieved by 2010 the Melbourne allocation will be supplemented with water that is dedicated for environmental (water-quality) purposes on the Goulburn and Broken Rivers. Assessments of regional social, economic and environmental impacts of this water-sharing arrangement have not been conducted. The Steering Committee has recommended in its Draft Report that despite the proposed pipeline bringing water to Melbourne, trade of water between farmers and city-dwellers will not be permitted. The rationale for this recommendation is not explained in the Draft Report.

The FBMP is largely focused on water-services infrastructure, although it is noted that the project will have significant economic (and therefore social) benefits in the region. For example, the Draft Report describes in considerable detail the various components of what the modernised water-delivery system will consist of, including interconnectivity, automated supply backbone, various connections to the supply backbone for different customers and customer irrigation systems. There are only brief references to any social benefits arising from having an efficient irrigation system and no references at all to how the modernised system might also facilitate delivery of environmental water to stressed rivers and degraded wetlands.

On paper, the water-sharing arrangements (75 billion litres each for irrigators, Melbourne and the environment) appear equitable and a reasonable justification of such significant public investment. Further, the annual allocation of these savings is to be equally divided between these user groups. Thus, if water availability is diminished by reduced inflows, the allocation will be adjusted appropriately and then distributed equally. However, the question of whether or not the environment will indeed receive its share remains. First, although there is a commitment that environmental water will have the same status as water available to Melbourne and to irrigators, the Minister still retains discretion over the allocation of this water.

Concerns remain therefore about the final outcomes for the environment and for rural communities of any movement of water out of the region. Whether it is sufficient to 'offset' such water by money flowing into the region through upgrading of irrigation infrastructure and some potential for provision of environmental water to stressed rivers returns us to the opening questions posed in this paper about the commodification and exchange value of water. Further, if the pipeline infrastructure is in place, will pressure mount for lessening the 'barriers to trade' to urban areas to permit private trading regimes? Potential price differentials for rural, as opposed to urban, water are a likely catalyst — especially with projected large price increases in metropolitan water in Melbourne that may emerge from any 'reform' of the metropolitan retail-water sector. These factors would seem to add to general privatisation pressures to create a regime of private property in water that does not distinguish rural from urban areas. If such a scheme was instituted, then the localised effects of 'risk-spreading' under any such potential regime is more problematic without firm and transparent legal safeguards for its operation to minimise its distributive-justice impacts. Alternatively, if we use a human-rights formula it may also raise dilemmas. Few would argue against the provision of basic water rights to urban dwellers in terms of drinking supply and basic hygiene on a basis guaranteed at law. Yet do urban dwellers have a right to 'social needs' water ahead of rural dwellers or the environment on the basis of efficient and highest-value use? Resolution of these complex questions will require rethinking how water is conceptualised, valued, managed and allocated within the Australian continent, including the respective processes of interaction between urban and rural areas.

Conclusion

The food-bowl modernisation example indicates that the process of legal and regulatory change in water resources is more diffuse than a simple imposition of 'market systems' to displace the existing elements of the state-based 'resource development' model (Bakker 2005: 546). Urban water-law reform in Australia is a regulatory sphere where many elements of state-based property, private property and market-efficiency models coexist. Further, the concepts, assumptions and terminology of resource economics and competition theory have been mediated through malleable definitions of the 'public interest' and competing definitions of water 'property' rights. Clearly, as the Victorian situation illustrates, these remain open-ended and responsive to further redefinition as the implementation of water-law reform proceeds. In the context of increased pressure for governments to respond to pressures of climate change and water security, it is critical that there be enhanced avenues for informed third-party comment and greater social debate about these questions of the public interest in water-resource development. Social science perspectives provide one such means of capturing a broader understanding of the social and cultural values of water that often seem to be either excluded or 'discounted' in the prevailing modes of analysis that inform urban water-supply planning and management. Law, as part of that wider perspective, can play a crucial role in shaping and implementing these normative perspectives.

In turn, adoption of such normative objectives will require associated statutory, institutional and regulatory adjustments, some of which may well entail potential readjustments of underpinning property constructs. The ongoing redefinition of what constitutes property in water as a state-based or private-law entity is inherent to reshaping the purposes of water regulation, the forms of governance and techniques of implementation within an overarching narrative of a move toward greater market ordering. Again there is a need for enhanced transparency about the assumptions upon which such decision-making proceeds rather than an implicit 'trust' in technological and market-based approaches. Indeed, policy solutions that are being proposed to remedy the water-security problem in urban areas are functionally constructed in prevailing analyses as market responses. Yet these 'market responses' still require explicit deployment of public supply-side 'technological' solutions (and legal and institutional support) to institute those efficiencies. Amalgamation of technological supply-side solutions with allocation/entitlement-oriented regimes is seductive. These approaches suggest the input of the public interest via consumer preference and individual-use values, while masking the extent to which demand-side options are displaced by supply-side security 'solutions'.

In conclusion, even if urban water governance is to be progressively moved within a private-property, market-based model, the hand of government in the market — provided it is much more transparent and not invisible — may well be appropriate to provide a counterpoint to the narrowness of private transactional formulations of property and efficiency that ignore third-party effects and environmental and social externalities. Third-party effects may be felt more acutely in non-urban contexts, even though the focus may be urban water provision. As Arnold (2005: 4) notes: '[J]ust because property is private, rather than public, does not mean that it is not subject to public controls and interest.' Alternatively, though, hybrid models of public/private regulation also need to transcend the accretion of earlier state-based technology-driven approaches in favour of more fine-grained regulatory forms that can include more complex understanding of the value of water in a long-term intergenerational sense, as well as appropriate and effective costing of externalities to address long-term water and environmental security.

References

The Age, 'The Trade in Rural Water to Supply Urban Swimming Pools', October 20, 2007.

Anderson, T. and Leal, T. 2001, *Free Market Environmentalism*, New York: Palgrave.

Arnold, C. 2005, 'Privatization of Water Services', *Pepperdine Law Review* 32: 561–604.

Ayers, I. and Braithwaite, J. 1992, *Responsive Regulation: transcending the deregulation debate*, Oxford: New York.

Bakker, K. 2000, 'Privatizing water, production scarcity: The Yorkshire drought of 1995', *Economic Geography* 76(1): 4–27.

Bakker, K. 2005, 'Neoliberalizing Nature? Market Environmentalism in Water Supply in England and Wales', *Annals of the Association of the American Geographers* 95(3): 542–65.

Black, J. 2002, 'Critical Reflections on Regulation', *Australian Journal of Legal Philosophy* 27: 1–35.

Bluemel, E. 2004, 'The Implications of Formulating a Human Right to Water', *Ecology Law Quarterly* 31(4): 957–1006.

Clark, S. 2003, 'The Murray-Darling Basin: Divided Power, Co-Operative Solutions', *ARELJ* 22: 67–80.

Connell, C. *et al.* 2007, 'Delivering the National Water Initiative: Institutional roles, responsibilities and capacities' in Hussey, K. and Dovers, S. (eds), *Managing Water for Australia: The Social and Institutional Challenges*, CSIRO Publishing: 127–40.

Connor, R. and Dovers, S. 'Property Rights Instruments: Transformative Policy Operations' in *Property: Rights and Responsibilities, Current Australian Thinking, Land and Water Australia*, available at: [http://www.lwa.gov.au/downloads/publications_pdf/PR020440.pdf]: 43.

Considine, M. 2005, 'Partnerships and Collaborative Advantages: Some Reflections on new forms of Network Governance', Centre for Public Policy, The University of Melbourne.

Cullen, P. *et al.* 2002, 'Blueprint for a Living Continent: A Way Forward from the Wentworth Group of Concerned Scientists', WWF Australia, available at: wwf.org.au>.

Department of Sustainability and Environment 2007, 'The Next Stage of the Government's Water Plan'.

Dryson, M. 2002, 'Using Existing Legislation to Recover and Protect Environmental Flows in the River Murray', in *Property Rights and Responsibilities: Current Australian Thinking*, Land and Water Australia, AGPS.

Dryzek, J. 1997, *The Politics of Earth: Environmental Discourses,* New York: Oxford University Press.

Eder, K. 1996, 'The Institutionalisation of Environmentalism; Ecological Discourse and the Second Transformation of The Public Sphere', in Lash, S. *et al.* (eds), *Risk, Environment & Modernity Towards a New Ecology*, Sage Publications.

Fisher, D. 2004, 'Rights of Property in Water: Confusion of Clarity', 21 *EPLJ*: 200–26.

Fisher, D. 2007, 'Delivering the National Water Initiative: the emergence of innovative legal doctrine' in Hussey, K. and Dovers, S. (eds), *Managing Water for Australia: The Social and Institutional Challenges,* CSIRO Publishing: 113–26.

Foester, A. 2007, 'Victoria's New Environmental Water Reserve: What's in a name?', *Australasian Journal of Natural Resources Law and Policy* 11(2): 145.

GHD Coliban Water/Goulburn–Murray Water, Goulburn-Campaspe Link Project Interim Report, May 2006.

Gleick, P. 1999, 'The Human Right to Water', *Water Policy* 1(5): 487–503.

Glennon, R. 2005, 'Water Scarcity, Marketing, and Privatization', *Texas Law Review* 83(7): 1873–1902.

Godden, L. 1997, '*Wik:* Feudalism, Capitalism and the State. A revision of land law in Australia?', *Australian Property Law Journal* 5(2–3): 162–80.

Godden, L. 2006, 'Efficiency and Environmental Sustainability: Using Market Mechanisms for Water Resources Regulation in Australia' in Barton, B., Ronne, A. and Hernandez L., *Regulating Natural Resources,* Oxford University Press: Chapter 8.

Grafton, R. Quentin and Ward M., 'Prices versus Rationing: Marshallian Surplus and Mandatory Water Restrictions', Crawford School of Economics and Government, The Australian National University; paper prepared for workshop Fenner School of Environment, ANU, 23–4 November 2007.

Gray, K. 1991, 'Property in Thin Air', *Cambridge Law Journal* 50: 252–307.

Heaney, A. *et al.* 2006, 'Third-party effects of water trading and potential policy responses', *Australian Journal of Agricultural and Resource Economics* 50: 277–93.

Hildering, A. 2005, *International Law, Sustainable Development and Water Management*, Euburon Publishers: Delft.

Hurlimann, A. 2007, 'Is recycled water use risky? An Urban Australian community's perspective', *Environmentalist* 27: 83–94.

Kellerman 2007, 'Ripple Effect', *Financial Review*, September 2007: 20.

Keohane, N. O., Revesz, R. and Stavins R. 1998, 'The Choice of Regulatory Instruments in Environmental Policy', *Harv. Envtl. L. Rev.* 22: 313–67.

Ladson, A. and Finlayson B. L. 2002, 'Rhetoric and reality in the allocation of water to the environment. A case study of the Goulburn River', *River Research and Applications* 18: 555–68.

Marsden, J. 2003, 'Water Entitlements and Property Rights: An Economic Perspective' in *Property: Rights and Responsibilities, Current Australian Thinking, Land and Water Australia*, available at: [http://www.lwa.gov.au/downloads/publications_pdf/PR020440.pdf]: 43.

Marsden Jacob Associates, 'Securing Australia's Urban Water Supplies: Opportunities and Impediments', Discussion Paper prepared for the Prime Minister and Cabinet, November 2006.

Miller, A. 2005, 'Blue Rush: Is an International Privatization Agreement a Viable Solution for Developing Countries in the face of an Impending World Water Crisis', *Indiana International and Comparative Law Review* 16: 217.

Office of State Owned Enterprises 1995, 'Reforming Victoria's Water Industry: The Restructured Metropolitan Industry'.

Parker, C. Scott, Lacey, N. and Braithwaite, J. (eds), *Regulating Law,* (1st edition) 2004: 1.

Picciotto, S. 2002, 'Introduction: Reconceptualising Regulation in the Era of Globalisation', *Journal of Law & Society* 29: 1–11.

Petrova, V. 2006, 'At the Frontiers of the Rush for Blue Gold: Water Privatization and the Human Right to Water', *Brooklyn Journal of International Law* 31: 577–613.

Posner, R. 1998, 'The Economic Theory of Property Rights: Static and Dynamic aspects' in R. Posner, *Economic Analysis of Law* (5th edition): 37.

Powell, J. M. 1989, 'Water Managers Ascendant' in J. M. Powell (ed.), *Watering the Garden State: Water, Land and Community in Victoria, 1834–1988* (1st edition).

Richardson, B. 2002, 'Environmental Regulation through Financial Institutions: New Pathways for Disseminating Environmental Policy', *Envtl. & Plan. L. J.* 19: 58–77.

Richardson, B. and Wood, S. (eds) 2006, *Environmental Law for Sustainability: a reader*, Oxford: Portland.

Scanlon J., Cassar, A. and Nemes, N. 2004, 'Water as a Human Right?', IUCN: Gland, Switzerland and Cambridge, UK.

Schneiderman, D. 2000, 'Constitutional Approaches to Privatization: An inquiry into the magnitude of neo-liberal constitutionalism', *Law and Contemporary Problems* 63: 83.

Smets, H. 2006, 'The Right to Water in National Legislation', available at: http://www.emwis.net/documents/database/semide/PDF/right-to-water_EN.

Sneddon, N. and Ellinghaus, M. (eds), *Cheshire & Fifoot's Law of Contract* (7th Australian).

Tarlock, D. 2004, 'Is there an Environmental Law?', *J. Land Use & Envtl. L,* 19: 213.

Teubner, G. 1983, 'Substantive and Reflexive Elements in Modern Law', *Law & Society Review* 17: 239–85.

Teubner, G. 1993, 'The "State" of Private Networks: The Emerging Legal regime of Polycorporatism in Germany', *Brigham Young University Law Review:* 553.

Victorian Auditor-General, Planning for Water Infrastructure in Victoria (April 2008), available at http://download.audit.vic.gov.au/files/Water_Infrastructure_Report.pdf>http://download.audit.vic.gov.au/files/Water_Infrastructure_Report.pdf

Victorian Competition and Efficiency Commission 2007, 'Inquiry into the reform of the metropolitan retail water sector'.

Vincent-Jones, P. 2002, 'Values and Purpose in Government: Centre-local Relations in Regulatory Perspective', *Journal of Law and Society* 29: 27–55.

WaterAid, see URL: [http://www.righttowater.org.uk/code/FAQs_2.asp#1]

Young, M, 2007, 'Linking Rural and Urban Water Systems' in Hussey, K. and Dovers, S. (eds), *Managing Water for Australia: The Social and Institutional Challenges,* CSIRO Publishing: 85–95.

Young, M., Proctor, W. and Quereshi, M. 2006, 'Without Water The economics of supplying water to 5 million more Australians', CSIRO/CoPS National Flagships Water for a Healthy Country.

Conclusion: A new solution

Patrick Troy

By the 1860s Australian cities were generally facing four problems with their water supplies:

- They had poor supplies of potable water, resulting in infections from water-borne contagions.
- They were unsanitary and had increasing difficulties, including threats to health, in dealing with the disposal of human and other wastes of urbanisation.
- They suffered periodically from poor drainage of stormwater.
- They experienced crises due to a lack of convenient supplies of water to fight fires.

While all four were important, the health of their population was the prime consideration in securing new water supplies. The mid-century recognition in England, documented by the sanitation reformer Edwin Chadwick, that many health problems were directly related to the lack of secure supplies of potable water (Flinn 1965) was followed by pressure in Australian colonies to develop such supplies (Dingle, this volume, Chapter 1).

Potable water

From settlement, Colonial administrations had tried to secure reliable supplies of potable water by exploiting sources 'beyond' the urban boundary, but growth of each colony was such that often the urban area quickly grew beyond the area reserved and supplies were compromised. Households harvested and stored rainfall from roofs in tanks and occasionally from surface runoff in underground cisterns. These supplies often failed in long summers or drought periods. In addition, it became increasingly apparent that underground cisterns provided poor-quality water because of infiltration of runoff and due to seepage of sewage effluent into the cistern (Lloyd et al. 1992). 'New' sources periodically had to be sought from further afield to provide the secure supplies of potable water.

As the cities grew, the demand for the reticulation of secure supplies of potable water increased. Underlying the development of these supplies was the assumption that the demand for water could always be met by seeking/developing new supplies. The initial assumption, in 1878, for demand in Newcastle, which was similar to Sydney and advised by the same engineers, was that personal consumption of 20 gallons (91 litres) per head per day was sufficient to meet the demands for consumption, food preparation and personal hygiene (33.2kL pa) but that this might rise to 50 to 80 gallons per head per day to meet the needs of manufacturing and garden watering (Lloyd *et al.* 1992). In Melbourne the estimated demand was originally 40 gallons per capita per day but reduced to 30 gallons per head per day before construction began (Dingle and Doyle 2003).

While potable water was needed for health reasons, the supply seemed reliable and generous enough to allow households to use water of a standard fit for domestic consumption for sanitation, to water gardens and for other uses. The seemingly adequate supply also meant that domestic bathing and laundry practices changed, with consequent dramatic increases in the discharge of wastewater from households. Households were no longer able to rely on the use of cess pits and drainage sumps to dispose of wastewater and 'excess' wastewater drained onto the street and into the general surface drainage system that was already inadequate to cope with stormwater runoff. The result was an increase in noisome flows of wastewater and sewage onto the city streets.

By 1880 the issue of managing waste disposal assumed greater proportion as urban populations grew. The worldwide popularity of Edwin Chadwick's ideas for improving urban sanitation and the development and increasing take-up of water closets exacerbated the sanitation problems in Sydney and Newcastle but also offered the idea for its solution in the form of the development of a piped sewerage system.

There was a neat symmetry in this. The supply of water met all the needs of households for potable water and there appeared to be water enough to provide the medium for the transport of wastes. This was seen as an elegant solution and in the original Chadwick proposal offered the first environmental solution to the management of human body wastes because it proposed to collect them and transport them to be used as fertiliser on nearby farmlands — a solution that, as Dingle points out in Chapter 1, was experimented with in Sydney and Adelaide but only seriously adopted in Australia by Melbourne.

The virtuous circle that was ultimately taken in most Australian cities was to develop a reticulated water supply and later a piped sewerage system to remove sewage. This solution was made more financially attractive for water authorities with the banning of rainwater tanks and the preferment of waste-management technologies that relied on water transport to the exclusion of technologies that did not. The attractions of water-based sewerage systems were so compelling that a networked sewerage system was developed to transport wastewater, human excreta and other wastes. This seemingly felicitous solution to the problem of sanitation ultimately led to a large environmental problem in the form of discharges of sewage to the ocean. Property owners were required to connect to the public water supply and later to the sewerage system on public-health grounds. Water consumption rose as households took advantage of the apparently abundant supplies for their flush toilets and for personal hygiene.

Stormwater drainage

Stormwater runoff became more problematic as the cities grew and more of their area was covered with impervious surfaces. The volumes of water were so great that it was infeasible to manage the runoff by using the sewerage system, so separate stormwater drainage systems were developed. They, too, drained directly into rivers, harbours, bays and oceans that abutted the cities and became significant sources of pollution of those waters.

The early decision to develop separate systems for sewerage and stormwater drainage meant that wastewater flows could avoid the peaking problems associated with storms — problems that would only be exacerbated as development of the cities led to the increasing coverage of drainage catchments by impervious surfaces. Although sufficient water falls as rain in the urban areas to meet their water requirements, this approach to stormwater management means that, even today, stormwater is discarded and treated as a 'problem'. How much of this is a consequence of the fact that surface drainage, as distinct from sewerage, was the responsibility of local government whereas in many cities sewerage was the responsibility of the water-supply authority has been little explored. It is interesting to note that in the recent drought some thought has been given to the possibility of redirecting stormwater, on occasion, to ensure that the sewers were properly flushed — the combined effect of water restrictions and water-saving behaviour having reduced sewage flows so that the systems had a tendency to block.

Twenty-first-century outcomes

The net effect of these nineteenth-century 'solutions' in the first decade of the twenty-first century is:

1. The per-capita consumption of water is now three times the level the original systems were designed to provide.
2. Stresses in the ecosystems from which water is abstracted to supply the cities.
3. Extreme stresses on the ecosystems into which wastewaters are discharged.
4. Stormwater runoff systems that are the major sources of pollution of the rivers, bays and harbours on which the cities are built.

The combined effect of rapid increase in population and massive increase in per-capita consumption meant that the demand for water soon outstripped supplies but the attraction and seeming felicity of the 'scientific' approach to water management fostered the engineering systems needed to increase supply — usually in the form of more dams which impounded the water in ecosystems further from the cities for transport to them. There was a comforting belief that there were always additional supplies available and all that was required was application of engineering skills to deliver them to the cities.

By the mid twentieth century, most Australian cities had exploited all the water resources available in their near hinterlands. Their supplies were in precarious balance with demand. Although they had originally been conceived of as 'health authorities', with a remit to protect the public health, water authorities had seen opportunities to avail themselves of the financial rewards arising out of increased consumption. The gradual acceptance by their residents of the commodification of water and their passive acceptance of increased use of water-using services and equipment increased consumption and brought with it increased financial rewards to the water authorities. The water authorities were remarkably efficient at harvesting, storing and transporting the available surface water resources, although system losses due to evaporation and leakages became increasingly important as they reached the limits of the 'natural supply'. As the population grew and was accompanied by increasing levels of per-capita consumption, the water authorities increasingly found themselves with few reserves to cope with vagaries in supply. By the end of the twentieth century the situation became critical, in part because the apparent reduction in long-run rainfall over dam catchments meant that reservoirs were operating with small reserves.

The response has been to seek ways of increasing supply and, as a temporary measure, to introduce water restrictions aimed particularly at reducing water consumption on uses outside the dwelling. Generally, water restrictions have had some success in reducing demand but the scale of reduction has not been large, with the notable exception of Brisbane where the Queensland Water Commission in 2007 restricted water consumption to 140 litres per person per day and managed to improve on the target within a very short period (although there is some expectation that once the new desalination plants, water recycling, new dams and water grid come into operation consumption will increase (QWC 2008) — possibly with encouragement of the water authorities to increase revenue). These measures have not allayed anxieties but the current drought has brought underlying problems in the management of national water resources into high relief and led the Commonwealth Government to initiate national water-policy reform.

Features of the demand and supply of urban water

Two aspects of the water system need to be borne in mind:

1. The demand for water has some seasonal variation, with summer demand being higher than winter, but the pattern of consumption is fairly constant

year on year for conventional housing and especially throughout the year for higher-density forms of housing (Troy *et al.* 2005).

2. The supply of water through the water catchments is highly variable, depending as it does on rainfall. This was less of a concern when the storage was large enough to allow for several years of consumption, but is now because the increase in population, together with the increase in per-capita consumption, produces a high and relatively constant demand while rainfall over the catchments in the larger cities appears to have declined.

Attitudes to personal hygiene and cleanliness practices had been changing since the Middle Ages (Vigarello 1988). Moreover, cultural and behavioural norms in domestic water use changed considerably in the developed world during the nineteenth century, all adding considerably to increased per-capita water use. This meant that people used flush toilets and flushed them with each use more often compared with earlier toilet practices. They also washed themselves more frequently. At first, this was by bathing, but this was replaced by the increasing popularity of showering, which led to greatly increased domestic water consumption (Shove 2002; Shove 2003 — this is a discussion of UK experience but accords well with Australian experience; see Davison, Chapter 3 of this volume) and wastewater generation. To some degree the popularity of showering is related to the pleasure of the act — especially once heated water was more readily available — as much as it was to notions of personal hygiene (Gilg and Barr 2006; Hand *et al.* 2003; Allon and Safoulis 2006; Safoulis 2006). As Davison, Chapter 3, shows, cultural and behavioural norms in domestic water use in Australia changed considerably over the last 150 years, all adding considerably to increased per-capita water use, especially in the cities.

A recent survey of Sydney households' attitudes (Troy and Randolph 2006) revealed their strong determination to maintain their level and nature of shower use. It also revealed considerable reluctance to reduce toilet flushing. These responses suggest that programs designed to reduce consumption from both activities may encounter strong passive resistance.

So we now have a paradox that water-supply systems, once determined by considerations of health and primary hygiene, are more driven by calculations of lifestyle that may well be counterproductive. For example, increased showering is claimed to have been accompanied by an increase in skin diseases (Shumack 2006).

Contemporary water consumption also both helps create the demand for, and is a consequence of, the form of development of the city. The traditional separate house in its own garden was (and remains) a strong expression of the felt needs of households for a degree of independence (Gaynor 2006). It may also create the possibility of a high level of food self-sufficiency.

Traditional housing not only provided the opportunity for a high level of domestic production (Mullins 1981a, 1981b and 1992) but it also 'explains' why it was such an effective cornerstone of the conservative philosophy expressed by Menzies in the 1940s and 1950s (Brett 1992) who successfully built on the desire of households for a home of their own with a small garden to gain and retain office nationally and to shape the policies which guided the massive growth of Australian cities in the 1950s and 1960s. Freestone (2000) documents how the garden-city movement shaped the nature of Australian cities, although Hall (2007) also documents the disappearance of gardens in the contemporary city.

This form of accommodation not only provided the opportunity for a high level of domestic production, it also 'explains' why the separate house and garden shaped the policies which guided the massive growth of Australian cities in the 1950s and 1960s. Paradoxically, this suburbanisation is seen by some as entrenching the resistance to reform of water-consumption practices.

The widespread take-up by households since the 1940s of washing machines (Davison, Chapter 3) led to increased water consumption. In earlier periods, washing clothes was a tedious affair. The advent of new machines offered to take some of the labour out of the washing task. The higher level of workforce participation by women and the increasing degree of consumerism since the 1940s was accompanied by a significant fall in the cost of clothing and manchester items in household budgets, which in turn meant that people were able to change their clothing and manchester items more often and meant that there were more clothes to wash. The ability of machines to wash clothing whenever it was convenient significantly increased water consumption. This take-up of water-using services and appliances that have been an integral part of Australian cities may now be seen also as entrenching resistance to reform of water-consumption practices.

Water consumption in the kitchen has also increased, although it remains a small proportion of internal household consumption. External consumption of water also increased with the increasing popularity of swimming pools and more recently of spas. Garden usage is also important, but in some cities it is less significant than might be assumed. In Sydney, for example, most households rely heavily on rainfall to maintain their gardens (Troy and Randolph 2006).

The point is not to lament these changes, but simply to appreciate their cumulative effects on values and expectations as well as on levels of consumption. And then to ask how such interdependencies might be untangled in the least obstructive, most efficient ways.

Development of water supply: The case of Sydney

Water services are provided by a government corporation and monopoly supplier (Sydney Water Corporation). It has a strong engineering culture overlain by a strong economistic approach to water-management issues. In the face of occasional criticism, it has developed a strong defensive institutional culture.

From its origins as a colony, Sydney has for the last two centuries responded by projecting the demand without any fundamental review of the services it provides and then setting out to provide the supply. Sydney Water Corporation's response to the increasing demand for water has been to follow the traditional 'project and provide' approach to water services.

It is now clear that Sydney cannot simply continue to harvest waters from sources outside its immediate region to meet what appears to be an unquenchable demand without serious environmental consequences and without failures in supply. This is acknowledged in the Metropolitan Water Plan, which was been developed in part to meet the increased demand for water from a predicted increase in Sydney's population of around one million over the next 25 years. The Plan implicitly assumes an ever-increasing supply to meet demand.

The focus on increasing supply of water in the 'traditional' way will eventually prove problematic and unmanageable because of the environmental stresses associated with the approach and, not least, on cost grounds. A more fruitful way of continuing to meet reasonable demands for potable water from Sydney Water Corporation's *existing* storage facilities such as Warragamba Dam lies in encouraging residents to reduce their water consumption and to accept greater responsibility for security of their own supply and wastewater management in a manner that reduces the demand for potable water, improves the sustainability of the city and simultaneously enables the government to meet new environmental targets.

There is an urgent need for a major change in the way demand for water should be managed at the level of the individual household, together with new measures to reduce the consumption of potable water in the home. Such an approach is built on the assumption that initiatives need to be taken to minimise the environmental stresses that accompany the present consumption of water and the management of wastewater flows (Guy *et al.* 2001). It is also assumes that we cannot simply turn to a new system and ignore the path-dependency effects of the water-supply and sewerage systems in place, a point made strongly by Dingle in Chapter 1. A new approach would need to be phased in as part of a new water-demand management model which would lead to less reliance on the traditional reticulation system and reduce the per-capita consumption of potable water supplied by Sydney Water Corporation.

Over the past six years we have had a plethora of national conferences and summits on water services. Much of the focus of these meetings has been on 'reform', which is code for privatisation of aspects of the services. The conferences have usually been built on the assumption that the present institutional structure of the urban water 'industry' is given, that the present engineering solutions are only to be made more efficient, that pricing is an acceptable, indeed major, tool for moderating demand (the shape of which is essentially taken as given) along with exploration of the privatisation of aspects of the water services to improve 'efficiency' but that there should be a vigorous search for 'new' sources. The program of the Sixth Annual 'Australian Water Summit' scheduled for March 2008, for example, provides a typical illustration of this constrained approach. These new sources tend to be ways of extracting more from existing dams by exploiting waters that were previously thought to be not worth using or were too costly to treat. Given the fact that all 'natural' sources are now fully exploited and in such a manner that there is little spare capacity to allow for the variation in rainfall and therefore of runoff from the dam catchments, the currently favoured solutions are to develop recycling plants (in the ACT these are coyly called 'water purification plants') and desalination plants.

We do not dwell here on the proposals to build desalination plants (although Peter Spearritt in Chapter 2 does document some of the consideration of such solutions to water supply in Southeast Queensland) which are now under construction in most major cities except to point out that desalination plants are not only expensive in environmental terms but they cannot easily be run efficiently in low-flow conditions (*SMH* 2007). The residents of Sydney are now being advised that the desalination plants now under construction will cost all households in Sydney an estimated $110 per year which, together with other measures the Sydney water corporation proposes, would increase water bills by $275 per year which is equivalent to an increase of 33 per cent in the average water bill (*SMH* 2007).

The present approach to water supply is to search for some 'new' source, with no review of the existing shape of the demand.

Reduction in demand for potable water

An alternative approach would be to explore the history of the way the consumption of water has changed to gain insights into possible avenues to reduce consumption. Davison's brief history here provides some valuable clues. In her exploration of the cultural determinants of water use, Lesley Head (Chapter 4) provides further evidence of ways in which consumption of potable water might be reshaped and reduced.

The problem of recycled water

One of the ways in which water corporations seek to increase supply is to sponsor the use of recycled sewage.

There is popular resistance to the human consumption of recycled water (Peter Spearritt Chapter 2; *Sydney Morning Herald* 2005b) because of anxieties about the efficiency of systems to eliminate the bacteria, protozoa and viruses commonly found in sewage, as well as the many biologically active molecules, such as drugs taken to control fertility, infection, hypertension, cholesterol, depression, and so on. The presence in sewage of preservatives added to food and beverages to which a significant minority, of people have allergic reactions is a further problem. A question also arises as to whether recycling systems can be maintained to produce high-quality water.

There is also increasing evidence that the engineering systems, including reverse osmosis, do not eliminate pharmaceutical drugs to a safe level, which may lead to increased health risks. Watkinson *et al.* (2007) report that 92 per cent of antibiotics are removed from treated sewage. A leading infectious-diseases physician and microbiologist, Professor Collignon, makes the point that this is only log 1 reduction, whereas for viruses and so on log 6 reductions are needed for microbiological safety (Collignon 2008).

One of the worrying features of the consideration over the use of recycled sewage in the potable-water supply is that this is being introduced without the benefit of community consultation. The community rejection of the proposal to use recycled sewage in the Towoomba water supply has led water authorities to proceed to develop such approaches to water without public plebiscite. Water-recycling plants are in operation, under construction or in advanced planning stages in Brisbane, Canberra and Sydney, for which there has been little or no independent research to explore the long-run health risks of such projects and no public discussion of them. The high energy consumption required to produce such water is rarely considered by the water authorities to be a serious problem.

The major preoccupation in the original development of urban water supplies was the supply of potable water to secure the health of the community. This was often reflected in the motto of the water supplier such as for Newcastle which proudly stated (in Latin) that it was 'For the Public Health' (Lloyd *et al.* 1992). Public-health concerns were also the rationale for making it compulsory for property owners to pay for the connection to the water supply and later to the sewerage system.

This single-mindedness was remarkably successful. The health of the community was improved dramatically.

Because only a small proportion of the water now consumed needs to be of potable quality, a significant proportion of the revenue of water authorities recently has been more related to consumption not primarily related to health needs. For a complex variety of reasons, water authorities have not been enthusiastic about pursuing strategies designed to reduce the reliance on the use of potable water for uses and activities that do not need such high-quality water. They have sometimes argued that it is too costly to develop dual-flow water systems or to develop methods to capture rainfall as alternative supplies. They have generally been more enthusiastic about the use of manufactured water from recycling or recycling sewage to provide supply and have tended to ignore the objections of the risks to community health raised by those opposed to the human consumption of recycled sewage.

Here we have a perverse situation. The original requirement to pay for water supplied to the property was justified on health grounds, yet those who now object on grounds of the risk to health by the forced consumption of recycled sewage are nonetheless required to pay for a service they regard as compromised.

While it should not be ignored, the large-scale recycling of wastewater for human consumption need not be part of a comprehensive solution to better urban water-conservation practice. If recycling wastewater for human use proves politically difficult, then there are alternatives.

Making the transition to sustainability

Few aspects of our approach to the development and management of cities have lasted 150 years. We no longer have the same building regulations. We have consigned the miasmatic theory of disease transmission that existed in Chadwick's time to the dustbin of history. We live and work in our cities in very different ways now than we did then, and we communicate with one another in ways that were unimaginable then. The way we consume energy and the forms of energy we consume today are very different from then. The governance of our cities is different now and we pay for a whole range of services in ways that were inconceivable then. Our concern for the environment demands a very different approach to the way we use natural resources now.

We accept that we live in a state of flux. Paradoxically, the path dependencies we have created in the water services we provide and the way we provide them is reflected not only in the technologies we use in consuming water services; it is also reflected in the cultures of the institutional and administrative arrangements we have devised for the management of water. This institutional culture has fed and been created by the 'predict and provide' approach which is taken. The preoccupation with pricing regimes as solutions to moderate demand does little to generate new thinking in approaches to water services. The present so-called crisis or 'water problem' may be an apposite time to review sanitation services and to develop a new approach that recognises our fundamental need for potable water to maintain our health standards and our need to manage human body wastes in a felicitous manner but one which minimises the use of water.

Even if it is acknowledged that present uses of water cannot be sustained and that the current approach to the water crisis by searching for ways of increasing supply is ultimately self-defeating, it would be impossible to arrange for a rapid transition from the way water services are currently provided. The 150 years of development of the water-supply and sewerage systems have shaped, and been shaped by, the development that has occurred in Australian cities. This creates a significant degree of path dependency in the way in which services are provided and must be taken into account in trying to find ways of continuing to provide a supply of potable water. A similar situation exists in relation to the provision of waste-management services.

While it is conceivable that alternative approaches to the provision of water-supply and waste-management services could lead to significant reductions in the consumption of water, any transition from the way these services are currently delivered must be pursued taking into account the rate of growth of the urban areas served and the rate of obsolescence of the existing reticulated services.

Currently, the additions to the built environment run at about 1–1.5 per cent per year, depending on the stage in the building cycle. By mandating all new developments to install rainwater tanks, greywater-recycling systems and dry-composting toilets would reduce the demand for potable water by up to 70 per cent per dwelling. By identifying areas where it would make sense to retrofit developments with such things, the rate of change of a new approach to water services could be doubled. Pursuing such a program for a decade would mean that after 10 years, 30 per cent of the urban development would be using 70 per cent less water per dwelling. Such savings would continue to be obtained as the older parts of the cities were progressively modernised. Similar savings could be achieved in all non-residential developments in the city. This would mean that the path-dependency effects of the present systems were recognised and taken into account as the city renewed itself. In the longer term, this would lead to a continuing and substantial reduction in demand for the publicly provided supply of potable water.

This suggests that changing the existing services may take some time and that several strategies may be pursued simultaneously.

The first would focus on an aggressive pursuit of efficiencies in the consumption and supply of water services in the existing urban development.

The second would be to require new additions to the urban stock to provide for the capture of rainwater runoff at the time of construction. Water so harvested could then be used to substitute for potable water supplied through the present reticulation system. Many current proposals include a requirement to plumb tanks to toilets and washing machines.

The third would focus on the development of a retrofit program to gradually change over the existing development, with the rate of change being dependent on the rate of obsolescence of the services.

This approach would minimise the problem of stranded assets, identified by Dovers in Chapter 5, which would be created if the rate of change to new systems was too rapid. The actual rate of change would be decided for different areas within the city following a detailed analysis of the water consumption in those areas and the efficacy of introducing new waste-management services and the costs of doing so. It would, of course, also explore the savings to be obtained from reducing water supplies and consumption and of reducing the management and treatment of waste flows.

Dovers also points out the need to explore changes in the institutional and regulatory arrangements currently employed in the management of water services generally. This seems to be the most difficult phase in developing a new approach to the solution of our water-services problem. Water authorities are simply loath to take a new approach. They take refuge behind economistic arguments that pricing structures can lead to reduced consumption but seem not to accept either the issues of rights of access to potable water or the equity aspects of the pricing regimes they favour. They also discount alternative approaches to supply, such as encouraging use of rainwater tanks or stormwater harvesting, on the grounds that they still have an obligation to provide water services in dry periods, arguing that the risk to services is too great. This leads them into arguments supporting the use of manufactured water in which the risks to health and to environmental stresses are heavily discounted. Their proposals also seek to avoid allowing the public any say in the decision-making.

An equitable pricing regime

A regulatory and retrofit strategy would need to be buttressed by a pricing policy which ensured that water was supplied to households at a minimum guaranteed volume per person per year at an equitable price. This might sensibly be seen as an inalienable environmental right for all residents. In this way, lower-income households and lower-consumption households would not be penalised. In Chapter 8, Godden explores issues of water rights, including the notion of environmental rights in water. The point here is that if the guaranteed level was close to the original design levels for most city services (approximately 30kl per person per year) and was the level needed to sustain life and provide for basic levels of personal hygiene, any consumption above that level could be considered as discretionary and priced to reflect that situation.

The price charged for consumption volumes above the minimum guaranteed volume could then be set at a rapidly escalating rate to ensure that those who used more than the minimum paid significantly more for water. This would mean that those with high external consumption would pay significantly more.

Such changes in the pricing regime would need to be accompanied by changes in the bill structure so that the 'fixed' charges component of water bills would be reduced to a minimum administrative charge or eliminated. This would ensure that a user-pays pricing system was properly directed at the higher water consumers. Such an approach would reduce the pressure in the current systems which effectively allow high consumers to impose financial costs or environmental consequences on others.

Constructing such a pricing regime would have the beneficial effect of relating water price to consumption in a way that was most likely to affect behaviour. It would also provide an incentive for the installation of water-harvesting facilities and lead to greater household water independence.

Present water consumption patterns

A recent ABS report revealed that in 2001, 25 per cent of water consumption in NSW was for outdoor or external purposes (Table C1). This was approximately the same as the proportion used in the bathroom (26 per cent) and for toilets (23 per cent). Kitchens and laundry uses accounted for the remaining 26 per cent. One of the paradoxes facing water managers is that although they have been successful in providing a reliable supply of drinking water, little of it is actually drunk (approx 1 per cent). The volume of potable water actually consumed, used in food preparation or cleaning cooking equipment and utensils, cutlery and crockery is about 10 per cent of household consumption.

Table C1: Average annual per-capita water consumption by location of use in 2001 (kL)

	NSW	VIC	QLD	SA	WA	ACT
Bathroom	26.3	26.5	26.0	18.5	22.4	18.7
Toilet	23.2	19.4	16.4	16.0	14.5	16.4
Laundry	16.2	15.3	13.7	16.0	18.5	11.7
Kitchen	10.0	5.1	12.3	12.3	10.6	5.8
Outdoor	25.3	35.7	69.0	62	66.0	64.4
Total	101	102	137	123	132	117

Derived from Tables 9.6 and 9.7 in ABS 2004

No regional breakdown of this consumption within NSW is offered in the ABS report (2004), but given that the great proportion of this consumption is accounted for by households in Sydney, the NSW figure can reasonably be taken as a close proxy for the Sydney Metropolitan Area at that time. Recent research shows that water restrictions on garden watering and car washing, the main targets of these restrictions, at best impacted on a minority of Sydney residents; namely, those who had gardens and bothered to water them, or those who regularly washed their cars at home. These turned out to be minority pursuits across households in Sydney, even before the introduction of restrictions.

The other key fact to note here about water consumption, as evidenced in several recent studies (IPART 2004a; Troy *et al.* 2005; Eardley *et al.* 2005), is that the size of household is a key determinant of domestic water consumption. A number of studies indicate that on a per-capita basis Sydney households in different forms of accommodation have, for all practical purposes, similar annual demand for water, at approximately 100kL (IPART 2004a & b; ABS 2004; Troy *et al.* 2005). Research also indicates that there were considerable economies of scale in domestic water consumption in Sydney. This implies that, per-capita water consumption is not dependent on the residential built form. Falling household size is likely to be accompanied by an increase in average per-capita consumption.

Given that the current restrictions on external water use have probably reduced such use as far as is possible, then it is only by reducing the consumption of potable water *inside* the home that real gains in winding back the growing demand for water services in Sydney can be made.

Whatever the cause of the increasing inability of the water-supply system to meet current demand, whether it is due to growth in demand exceeding the supply, the need to maintain environmental flows, reduced runoff in the dam catchments due to long-run climatic cycles or to global climate change, there is an urgent need to re-examine Australian cities' water-services systems. This is needed to make the cities more water independent without at the same time creating unacceptable stresses on the regions from which water is abstracted or of creating environmental stresses in the water bodies around them into which wastewaters are discharged.

City water corporations have undertaken major exercises in demand management which significantly reduced consumption, most of which has been achieved mostly through improved efficiency in commercial and industrial activities. Mandatory restrictions on domestic water consumption, with severe penalties for those breaking the restrictions, have also been used to reduce consumption. The totality of these measures, however, remains insufficient to be able to rely on dams as the major supply.

A variety of alternative sources of water have been proposed in each city. In Sydney these include increased extraction from the Shoalhaven River south of Sydney (a river which is already stressed), large-scale recycling, extraction from aquifers in the Sydney region and building a major desalination plant. All proposals also imply continuation of the nineteenth-century solution to meet the demand for water by increasing supply. Before adopting any of these 'solutions' it would be apposite to review the nineteenth-century decision-making to try to understand how Sydney has reached the current crisis and to explore alternative methods of providing essential water services. The same could be said of other large cities such as Brisbane where a significant proportion of the consumption of water is used for cooling water flows in power stations and where the basic demands for water have not been reviewed. The draft water plan for SE Queensland (QWC 2008) nonetheless proposes five desalination plants and recycling of sewage (without public consultation) to maintain water services. In the case of the desalination plants, little consideration appears to have been given to the increase in greenhouse gases and therefore increased climate change pressures to which their operation would lead.

Many of the proposals to increase supply by exploiting wastewater flows assume that the waste streams are available to be sold. However, in Chapter 7 Gray and Gardner advance a compelling argument that waste may be seen as belonging to those who create it. This raises several considerations, including whether or not those discharging body wastes can object to their wastes being sold to others without their approval or of being used in recycling systems without their approval.

The development of reticulated water supply and sewerage systems in Australian cities in the late nineteenth century led to improved personal hygiene and improved sanitation, which was reflected in dramatic improvements in the health of communities. This success has coloured the approaches to water supply and management ever since and has raised community expectations that the present water-services systems can continue to do so. Unfortunately, they cannot.

Rather than simply increasing supply, a different strategy is now required to significantly reduce consumption of potable water. The strategy must acknowledge that the need to supply potable water for drinking and basic health reasons remains but that for other purposes individual and community expectations have to change. The question is: how can this be achieved at the same time as the use of potable water for purposes and activities that do not need to use water of drinking quality is reduced in an equitable manner? The current 'drought' provides the need for short-term measures to begin the process of re-educating people, of changing their patterns of consumption, of reshaping some of their behaviour and attitudes. The increasing acceptance of the reality of climate change and, with it, the increased variability in rainfall provides opportunities to change expectations and cultural norms that affect the patterns of consumption in potentially a more profound way.

Two basic approaches suggest themselves:

1. Measures to reduce consumption of potable water and encourage consumers to accept some responsibility for their consumption by making use of locally available water resources.

    Two possible sources suggest themselves:
    A. Rainwater tanks

        Rainwater tanks were, until the 1890s, the most common supply for most city households. They were made illegal in many cities (for example, Sydney and Newcastle) to ensure the financial viability for the then developing water-supply authorities and until recently were not allowed to be plumbed into the interior of dwellings. They were also banned because of alleged health risks. Whatever the justification for the position taken then, the current situation is that it is now possible to discard the first rainfall to flush the roof clean, ensuring that contamination of the tank water by bird and animal droppings is negligible. The second health argument was that tank water had high levels of lead in it. This was alleged to be from the lead flashing used in roofing and from the lead paints used. Neither has been allowed for many years, so this cannot be a major source of contamination now. While it is possible that lead may be above acceptable limits in tanks harvesting water from older housing, the use of 'first-flush bypass' systems greatly reduces the risk of pollution from lead and other heavy metals. The banning of lead additives in petrol also eliminated the possibility of lead being 'washed' into storage tanks through rainfall.
    B. Recycling and storage of treated of greywater

        Greywater cannot be stored for long before it becomes a nuisance and even a health risk. It is possible to treat the greywater on site for on-site uses such as toilet flushing, laundry and gardening. Apart from the use for gardening, this means that greywater should be treated and stored. The 'production' of greywater is slightly more than that used for toilet flushing so that the volume to be stored should be sufficient to maintain toilet flushing for a few days. Demand for garden watering may not be as consistent as that for toilet flushing so that the tank for storing treated greywater may need to be large enough to hold water for both activities.

2. Employment of technologies that enable the community to maintain sanitation objectives and meet its ambitions of comfort and convenience without consumption of potable water.

    Currently about a quarter of the per-capita average annual consumption of potable water is used to clear the toilet basin but this is not sufficient to transport the approximately 500 kilos of urine, faeces and paper 'produced' per capita annually (note that this is about twice the amount estimated to have been 'produced' at the time of the Chadwick report on sanitation. One of the problems was that while the Chadwickian solution to sanitation was based on water-borne transportation of wastes it assumed a relatively low flow of water. The sewers themselves had relatively high gradients.

Changes in water-using behaviour increased the wastewater flows, which had two consequences.

The first was that with higher flows the sewerage lines could be laid at lower gradients, leading to significant economies in the construction and operation of sewerage systems. One consequence of this approach is that currently the sewers need water in addition to that required to flush toilets to transport wastes to the treatment plants. This means that there is a tendency for sewerage-system managers to be less enthusiastic about measures to reduce 'wastewater' discharges to the sewer. The problem of dwellings 'producing' less sewage and wastewater flows on average thus lead to tendencies for sewers to 'block'. This has necessitated the release of potable water directly to the sewers to ensure that the system functions. Examples of this 'problem' may be seen in holiday and retirement areas that have rapidly expanded or where only small proportions of the dwellings serviced by a sewerage system are occupied at one time.

The second was that sewage treatment plants are required to treat ever-increasing volumes of water to increasing standards to minimise the environmental stresses from the urine, faeces and other wastes discharged to the sewer. The fact that approximately 40 per cent of the potable water delivered to dwellings (some of it being bathroom and laundry discharge) is now used to transport toilet wastes should itself cause questions to be raised about the efficacy of the present approach to sanitation services.

Other waste-management systems

The preference for water-borne sewerage systems meant that little encouragement has been given to other methods of managing human body wastes. There is a wide variety of approaches to provision of waterless or dry-composting toilets, with many systems developed in Sweden and the United States. Some of these separate the urine flow and many require low-powered air venting of the composting chamber. A number of similar systems using different ways of managing or recovering the compost have been developed in Australia. The system developed for medium density housing in Melbourne (GHD 2003) shows that such systems could be developed for urban areas and achieve very significant savings in water use. Using such a system would mean a saving of about 19 per cent water in Melbourne. Use of dry-composting toilets would mean that the recycled greywater would be available to maintain gardens and for laundry use. The use of dry-composting toilets would not only reduce water consumption but would enable the recovery of the composted material for use as garden or farm fertiliser. Use of a dry-composting system that enabled separate collection of urine flows would not only simplify the composting process but would provide a supply for processing to higher-grade 'natural' fertiliser or, suitably diluted, could be used in house gardens. It is important to acknowledge that today's dry-composting toilets are very different from the earlier manual systems for removal of sewage.

Households are under minimal pressure to reduce their consumption or to desist from discharging difficult or dangerous material to the sewage stream which complicates or makes difficult the operation of sewage treatment systems. Moreover, the structure of water pricing gives households little incentive to reduce their consumption.

Providing a subsidy and/or mandating the installation of low- or no-water toilets in all new developments would quickly reduce consumption of potable water substantially.

Local-area water management

Other ways of indirectly reducing demand for potable water include developing local-area harvesting of stormwater runoff for use in public parks and gardens. This will become more important as climate change proceeds because it will be even more important to encourage and support the growth of trees and shrubs to help manage the production of CO_2 emissions. More energetic encouragement of the use of greywater for the maintenance of trees and shrubs around dwellings would have a similar effect.

Separating the water-supply services from the sanitation services would lead to significant reductions in water consumption and sewage flows. The new approach would require dwellings to install a rainwater tank, a greywater-recycling system and a storage tank for the treated greywater. These components would increase the cost of dwellings but there would be significant savings in the dwellings' plumbing and in their water-supply system.

Installation of a new waste-management system would lead to significant reductions in sewage flows, which would lead to economies in the development and operation of the sewerage system. Reduction in the sewage discharge from dwellings would lead to smaller volumes requiring to be treated at sewage treatment plants and, in turn, smaller volumes to be discharged into receiving ecosystems.

The installation of dry-composting toilets would greatly reduce the need for sewerage services, as well as reducing water consumption. Such toilets would be cheaper to install than the present water-based flushing systems; moreover they would greatly reduce the environmental stresses currently experienced in the water bodies into which sewage is discharged.

The significant savings in the water-supply, sewerage and stormwater-management systems could be used to subsidise the installation of the new approach to water services. The reduction in the volume of potable water supplied by the water-supply network would leave more water to be applied to maintain environmental flows and to provide a more secure supply in dry periods.

Securing a similar degree of water independence for households in multi-unit developments would, in principle, be no different from those in traditional housing, although the collection of rainwater and the processing and storage of recycled water would present slightly different challenges.

As discussed above, water authorities were originally created as public-health agencies and they have been successful in that mission. Water consumption is now more a result of the commodification of water and changes in behaviour. One consequence of the need to encourage more local responsibility for water services is that the health standards of the community would need to be protected. Local water storage and waste management would need to be regulated and compliance with the regulations checked regularly to ensure that the water supply was of a high standard. Checking the quality of household and other supplies could become part of the obligations of the meter readers who visit consumers monthly to record their consumption.

Industrial and commercial development

A similar approach to the water supplied to new and existing industrial and commercial undertakings would also reduce the demand on potable water supplies and lead to similar economies in the water-supply, sewerage and stormwater-management systems.

Households, industry, commerce and public facilities would use significantly less potable water, which in turn would mean that the construction of new storage and large-scale treatment plants could be delayed, possibly indefinitely. There would be less need for high-volume reticulation of water-supply systems or for sewerage system and treatment plants. A major benefit would be that households and industrial and commercial undertakings would become more responsible for managing their own affairs.

An additional benefit would be that the stormwater runoff problem would be reduced, which in turn would reduce the pollution load in the rivers, harbours and bays on which Australian cities are built. The water-supply system would also be less vulnerable to attack or other disruption.

The reduced stormwater runoff could also be captured for treatment and recycling for industrial use, as well as for irrigation of public parks and gardens. It could also be used to maintain the environmental flows in rivers and other water bodies. Capturing and treating the reduced stormwater runoff would lead to reduction in the environmental stresses currently experienced by coastal and river waters into which untreated stormwater currently drains.

The nature of the water-supply services would change from one focused on large-scale catchment management to a much more localised set of catchments operated in a quasi-cascade form. Using the water resources on each block for the developments on them would not only ensure that residents and businesses became more aware of, and responsible for, their own supply as much as possible, it would also ensure that the provision of local water services for parks and public gardens made better use of the local drainage flows, including stormwater runoff. In this way, the present problem of the pollution of the cities' bays, rivers and harbours would be greatly reduced.

Institutional arrangements

The adoption of the approach outlined here would require changes to existing institutional arrangements. The first step would be to revise the regulations governing installation of rainwater tanks, waterless toilets or greywater treatment and recycling systems. Such institutional revision would also enable households to refrain from connecting to the sewerage system.

The present health regulations governing rainwater tanks, dry-composting toilets and greywater-recycling systems would need to be reviewed. Clearly, health objectives need to be secured but innovations in these technologies need to be recognised and improvements acknowledged in revised regulations controlling the installation and management of such systems. The powers of local government authorities would need to be revised to enable them to approve developments using modern water services and sanitation facilities.

City water corporations might sensibly be able to revert to their original role as a 'health authorities'. This would resolve the conundrum created by the enthusiastic adoption of the Chadwickian approach to the supply of potable water and the provision of sanitation services.

Conclusion

The point has been reached where it would be timely to reconsider the water services supplied to dwellings in Australian cities. It would also be timely to reconsider the ways in which waste-management services are provided. The situation facing all cities in Australia is that the water used to maintain their sewerage systems now accounts for almost half the water consumed inside the dwelling. This is putting the cart before the horse. The failure to reconsider the present water-supply and waste-management systems is leading to a moral panic in desperate searches for 'new' sources of water. All the options for these 'new' sources of water are expensive and environmentally damaging. The cities would be better served if more attention was paid first to ways of reshaping the demand for potable water and secondly reconsidering the ways in which wastes are managed.

Chadwick had, through his work on the Poor Law Commission, insisted on evidence in challenging the conventional wisdom of his time. His empirical research and that of others was based on the assumption that there was an inexhaustible supply of water. It was also based on the understanding that households consumed small volumes of water for all their wants. Nor did he or any of his colleagues understand that the great increases in the urban populations, partly as a result of the effectiveness of his reforms of sanitation, would lead to the burgeoning cities that followed.

References

Allon, F. and Sofoulis, Z. 2006, 'Everyday Water: cultures in transition', *Australian Geographer* 37/1: 45–55

Australian Bureau of Statistics 2004, *Water Account Australia 2000–01*, Catalogue No. 4610.0, Canberra.

BASIX Building Sustainability Index 2005, Department of Planning NSW. Available at: http://www.basix.nsw.gov.au/information/about.jsp. Downloaded 12 April 2007.

Brett, J. 1993, *Robert Menzies' Forgotten People*, Pan Macmillan: Sydney.

Collignon, P. 2008 — Professor, Infectious Diseases, Physician and Microbiologist, Director Infectious Diseases Unit and Microbiology Department, The Canberra Hospital. Professor, School of Clinical Medicine, Australian National University. Personal communication.

Dingle, T. and Doyle, H. 2003, 'Yan Yean. A history of Melbourne's early water supply', Melbourne.

DIPNR 2004, *see* NSW Department of Infrastructure, Planning and Natural Resources.

Eardley, T., Parolin, B. and Norris, K. 2005, 'The Social and Spatial; Correlates of Water Use in the Sydney Region', Final Report of the research project for the Water Research Alliance, University of Western Sydney.

Flinn, M. W. (ed.) 1965, *The Sanitary Condition of the Labouring Population of Great Britain by Edwin Chadwick, 1842*, Edinburgh University Press.

Freestone, R. 2000, 'Planning, Housing, Gardening: Home as a Garden Suburb' in Troy, P., *European Housing in Australia*, Cambridge University Press.

Gaynor, A. 2006, *The Harvest of the Suburbs: An environmental history of growing food in Australian Cities*, University of Western Australia Press: Perth.

Gilg, A. and Barr, S. 2006, 'Behavioural Attitudes toward water saving? Evidence from a study of environmental actions', *Ecological Economics* 57: 400–14.

Guy, S., Marvin, S. and Moss, T. 2001, *Urban Infrastructure in Transition: networks, Buildings and Plans*, Earthscan: London.

Gutteridge Haskin and Davey (GHD) 2003, Composting Toilet Demonstration Feasibility Studies, December.

Hand, M., Southerton, D. and Shove, E. 2003, 'Explaining Daily Showering: A Discussion of Policy and Practice', Working Paper series No. 4 Economic and Social Science Research Council, Sustainable Technologies Programme.

Hall, T. 2007, 'Where have all the gardens gone? An Investigation into the Disappearance of Back Yards in the Newer Australian Suburbs', Issues paper, Research Series, Urban Research Program, Griffith University. Also in Hall, T. 2008, 'Where Have All the Gardens Gone?', *Australian Planner* Vol. 45, No. 1: 30–7; and Hall, T. 2008, 'A Cautionary Tale', *Town and Country Planning,* Vol. 77, No. 2: 98–100.

IPART 2004a, 'Residential Water Use in Sydney, the Blue Mountains and Illawarra', Research Paper No.26, Independent Pricing and Regulatory Tribunal of NSW, Sydney.

IPART 2004b, 'The Determinants of Urban Residential Water Demand in Sydney, The Blue Mountains and Illawarra', Working Paper No.1, Independent Pricing and Regulatory Tribunal of NSW, Sydney.

Lloyd, C., Troy, P. and Schreiner, S. 1992, *For the Public Health: The Hunter District Water Board 1892–1992*, Longman Cheshire: Melbourne.

Melosi, M. 2000, *The Sanitary City: Urban Infrastructure in America from Colonial Times to the Present*, Johns Hopkins University Press: Baltimore.

Morgan, B. 2004, 'Water; frontiers markets and cosmopolitan activism', *Soundings: a Journal of Politics and Culture*, 28: 10–24.

Morgan, B. 2006, 'Turning off the tap, Urban Water Service Delivery and the social Construction of Global Administrative Law', *European Journal of International Law* 17: 215–47.

Mullins, P. 1981a, 'Theoretical perspectives on Australian Urbanisation. I: material components in the reproduction of Australian labour power', *Australian and New Zealand Journal of Sociology* 17(1): 65–76.

Mullins, P. 1981b, 'Theoretical perspectives on Australian Urbanisation, II: social components in the reproduction of Australian labour power', *Australian and New Zealand Journal of Sociology* 17(3): 35–43.

Mullins, P. 1996, 'Households, Consumerism and Metropolitan Development' in Troy, P. (ed.), *Australian Cities: Issues, Strategies and Policies for the 1990s*, Melbourne, Cambridge: 87–109.

NSW Department of Infrastructure, Planning and Natural Resources (DIPNR) 2004, 'Meeting the Challenges: Securing Sydney's water future', *The Metropolitan Water Plan 2004*, DIPNR, Sydney.

QWC 2008, *Water for Today, Water for Tomorrow*, Queensland Water Commission, Brisbane.

Shove, E. 2002, *Converging Conventions of Comfort, Cleanliness and Convenience*, Dept. of Sociology, Lancaster University, Lancaster, UK, at: http://www.comp.lancs.ac.uk/sociology/papers/Shove-Converging-Conventions.pdf

Shove, E. 2003, *Comfort, cleanliness and convenience: the social organization of normality*, Berg: Oxford and New York.

Shumack, S. 2006, Personal Communication.

Sofoulis, Z. 2005, 'Big water, everyday water: a socio-technical perspective', *Journal of Media and Cultural Studies* 19(a): 407–24.

The Sydney Morning Herald 2007, 18 September: 1. http://www.sydneywater.com.au/SavingWater/InYourHome/WaterFix. Downloaded 12 April 2007.

Troy, P., Holloway, D. and Randolph, B. 2005, *Water Use and the Built Environment: Patterns of Water Consumption in Sydney*, City Futures Research Report No.1, City Futures Research Centre, Faculty of Built Environment, UNSW and Centre for Resource and Environmental Studies, ANU.

Troy, P. and Randolph, B. 2006, 'Water Consumption and the Built Environment: A Social and Behavioural Analysis', Research Paper No. 5, City Futures Research Centre, Faculty of the Built Environment, University of New South Wales. This paper can be downloaded from www.cityfutures.net.au

Turner, A., White, S., Beatty, K. and Gregory, A. 2005, 'Results of the Largest Residential Demand Management Program in Australia', presented at International Conference on the Efficient use and Management of Urban Water, Santiago, Chile.

Vigarello, G. 1988, *Concepts of Cleanliness: Changing attitudes in France since the Middle Ages*, Cambridge University Press: Cambridge

Watkinson A. J., Murby, E. J. and Costanzo, S. D. 2007, 'Removal of antibiotics in conventional and advanced wastewater treatment: Implications for environmental discharge and wastewater recycling', water research,

National Research Centre for Environmental Toxicology, 39 Kessels Road, Coopers Plains, Brisbane, Qld 4108, Australia; Cooperative Research Centre for Water Quality and Treatment, PMB 3, Salisbury, SA 5108, Australia.

Index

www.ingramcontent.com/pod-product-compliance
Lightning Source LLC
Chambersburg PA
CBHW061227270326

41928CB00025B/3425